Metamaterials

Contents

CHAPTER 1

Introduction to Metamaterials

1.1 DEFINITION AND CONCEPT OF METAMATERIALS

1.1.1 Definition

Metamaterials are artificially engineered materials designed to have properties not usually found in naturally occurring substances. Their unique behavior arises from their structure rather than their chemical composition. Typically, metamaterials are constructed from periodic or patterned arrangements of smaller elements, often at scales smaller than the wavelength of the external stimuli they interact with, such as electromagnetic waves, sound waves, or mechanical vibrations.

1.1.2 Concept

The concept of metamaterials is based on manipulating waves—electromagnetic, acoustic, or elastic—in ways that natural materials cannot. Instead of relying on atomic or molecular properties, metamaterials derive their unusual characteristics from their engineered microstructure. By carefully designing the shape, size, geometry, and orientation of these structural units, scientists can control the propagation of waves through the medium.

For example, electromagnetic metamaterials can bend light in unusual ways, enabling phenomena like negative refractive index, cloaking (rendering objects invisible to certain frequencies), and superlensing (creating images beyond the diffraction limit). Similarly, acoustic metamaterials can control sound waves for applications like noise reduction or soundproofing, while mechanical metamaterials can manipulate stress and strain in structures.

In essence, metamaterials represent a new class of materials where function is determined by structure rather than composition, making them a groundbreaking advancement in material science with wide-ranging applications in optics, telecommunications, defense, medicine, and engineering.

1.2 HISTORICAL DEVELOPMENT AND BREAKTHROUGHS

The historical development of metamaterials is a story that bridges centuries of scientific discovery, beginning with early explorations of light and wave theory, advancing through the theoretical predictions of Veselago, and culminating in experimental breakthroughs at the turn of the twenty-first century. From the first demonstrations of negative refraction and superlensing to the captivating concept of invisibility cloaks and the practical design of metasurfaces, metamaterials have reshaped how scientists think about material properties and wave manipulation. The journey reflects the convergence of physics, engineering, and nanotechnology, producing a field that continues to evolve with new theoretical insights and technological advances. As research progresses, metamaterials promise to influence diverse industries and redefine the limits of what materials can achieve, making them one of the most exciting scientific breakthroughs of modern times.

1.2.1 Early Foundations of Wave Manipulation

The historical development of metamaterials is deeply tied to humanity's longstanding fascination with the control of waves, especially light and sound. Long before the term metamaterial was coined, natural philosophers and scientists pondered how waves could be manipulated to produce unusual effects. Early optics research during the Renaissance by scholars such as Johannes Kepler and René Descartes laid the groundwork for understanding refraction, reflection, and the behavior of light in different media. With the advancement of physics in the seventeenth century through Isaac Newton's work on light and Christiaan Huygens' wave theory, the conceptual foundation for wave manipulation was strengthened. These early efforts did not directly lead to metamaterials but established the essential scientific knowledge of wave properties that later inspired the creation of artificial structures designed to control them.

1.2.2 Nineteenth-Century Theoretical Background

The nineteenth century witnessed a series of breakthroughs that indirectly contributed to the metamaterial concept. James Clerk Maxwell's formulation of electromagnetic theory in the 1860s was a critical turning point, as it provided a unified description of electricity, magnetism, and light. Maxwell's equations predicted that electromagnetic waves could propagate through space and interact with materials, laying the basis for future innovations in engineered structures. At the same time, scientists like Heinrich Hertz confirmed the existence of electromagnetic waves experimentally, which provided tangible evidence of the concepts predicted by theory. These discoveries highlighted the relationship between material properties and wave behavior and encouraged later generations to consider the possibility of designing artificial materials to manipulate electromagnetic fields in ways natural substances could not.

1.2.3 Early Twentieth-Century Speculations

The early twentieth century brought new curiosity about materials with exotic properties. In 1904, the Russian scientist Victor Veselago theoretically described the possibility of materials with a negative refractive index, predicting that such media could bend light in the opposite direction compared to ordinary materials. His work, published in 1968, was not immediately recognized because no such materials were known to exist at the time. However, Veselago's paper became a cornerstone for later developments in metamaterials. He outlined the potential of materials with simultaneously negative permittivity and permeability, suggesting that such substances would exhibit unprecedented optical properties such as reverse Snell's law and backward wave propagation. Although this idea remained theoretical for decades, it provided a roadmap for experimental breakthroughs that came much later.

1.2.4 Rise of Artificial Structures in Microwave Engineering

The mid-twentieth century saw the emergence of artificial structures used in microwave engineering, which foreshadowed the development of metamaterials. Engineers experimented with metallic inclusions and periodic arrangements to create materials that could manipulate microwaves for radar and communication technologies. For example, frequency selective surfaces were developed to control the transmission and reflection of electromagnetic waves at particular frequencies. These engineered surfaces revealed that material properties could be tailored through structure rather than composition, echoing the same principle that would later define metamaterials. The growing interest in artificial dielectrics and engineered composites demonstrated the potential of man-made structures to extend beyond the limitations of natural materials.

1.2.5 Birth of Modern Metamaterials

The true breakthrough in metamaterial research occurred at the end of the twentieth century when scientists began to design and experimentally verify materials with properties predicted by Veselago decades earlier. In 1999, Sir John Pendry of Imperial College London proposed a practical design for negative permeability using arrays of split-ring resonators. His work showed that electromagnetic fields could be manipulated by carefully engineered metallic structures at scales smaller than the wavelength of radiation. This was followed by the creation of the first left-handed material in 2000 by David Smith and his team at Duke University, who combined split-ring resonators with metallic wires to demonstrate a negative refractive index at microwave frequencies. This achievement was the first experimental proof of Veselago's theoretical predictions and is often regarded as the formal birth of metamaterials as a scientific field.

1.2.6 Breakthroughs in Negative Refraction and Superlensing

The confirmation of negative refraction marked a turning point that attracted enormous interest from the scientific community. One of the most important breakthroughs was the concept of the superlens, proposed by Pendry in 2000.

He suggested that a metamaterial with negative refractive index could overcome the diffraction limit of conventional lenses and focus light beyond the scale of wavelengths. This idea promised revolutionary advances in imaging technologies, such as the ability to visualize nanoscale structures and biological samples with unprecedented clarity. Experiments in subsequent years confirmed aspects of this prediction, fueling excitement about the practical potential of metamaterials in optics and beyond.

1.2.7 Advancements in Optical and Acoustic Metamaterials

While initial breakthroughs occurred in the microwave regime, researchers quickly extended their efforts to higher frequencies, including the visible spectrum. This required the miniaturization of metamaterial structures to the nanoscale, which became possible due to advances in nanofabrication techniques. Optical metamaterials emerged, enabling control of visible light in ways never before achievable. At the same time, researchers adapted metamaterial concepts to sound and mechanical vibrations, giving rise to acoustic and elastic metamaterials. These innovations allowed scientists to manipulate sound waves to create acoustic cloaks, perfect sound absorbers, and noise-canceling devices. The extension of metamaterial principles to multiple domains underscored the universality of the concept and expanded its applications.

1.2.8 Vision of Invisibility Cloaks

One of the most captivating breakthroughs associated with metamaterials was the concept of cloaking. In 2006, researchers from Duke University demonstrated the first experimental electromagnetic cloak that could guide microwaves around an object, rendering it effectively invisible to detection at those frequencies. This achievement was inspired by the theory of transformation optics, which uses metamaterials to bend light paths in ways that mimic curved space. The demonstration of cloaking sparked both public fascination and intense scientific interest, raising the possibility of advanced stealth technologies, medical imaging applications, and novel optical devices. Although practical invisibility cloaks for visible light remain a challenge, the proof-of-concept experiments marked a milestone in metamaterial research.

1.2.9 Expansion into Metasurfaces and Two-Dimensional Structures

As research evolved, scientists realized that similar wave-manipulating effects could be achieved using thin, two-dimensional structures called metasurfaces. These structures, typically only a fraction of a wavelength thick, offered a more practical and efficient way to control electromagnetic waves compared to bulky three-dimensional metamaterials. Metasurfaces enabled functionalities such as holography, beam steering, and polarization control with simpler fabrication requirements. The development of metasurfaces represented a key breakthrough that broadened the scope of applications from communication systems to compact imaging devices, wearable sensors, and photonic chips.

1.2.10 Twenty-First-Century Innovations and Emerging Applications

In the twenty-first century, metamaterial research has diversified into many branches, leading to applications in fields as varied as wireless communication, medical diagnostics, energy harvesting, and defense. One important development is the integration of metamaterials with tunable and reconfigurable elements, such as liquid crystals, graphene, and microelectromechanical systems, which allow dynamic control over wave propagation.

These adaptive metamaterials are being explored for use in advanced antennas, 6G wireless networks, and real-time imaging systems. Another breakthrough is the development of topological metamaterials, inspired by concepts from condensed matter physics, which can guide waves along surfaces without loss or scattering. Such innovations represent a new frontier in the study of structured materials and open the door to practical, robust devices.

1.3 NATURAL VS. ARTIFICIAL MATERIALS

The distinction between natural and artificial materials reflects not only their origin but also the trajectory of human progress. Natural materials, shaped by nature, provided the foundation for early civilizations, while artificial materials, crafted by human ingenuity, have expanded the boundaries of what is possible. Each has unique strengths, whether it is the ecological compatibility of natural substances or the precision and versatility of artificial ones. Yet both categories remain indispensable, often complementing one another in applications that balance performance with sustainability. As societies face new challenges in technology and environmental responsibility, the relationship between natural and artificial materials will continue to evolve, guided by the quest for innovation and the need for harmony with nature.

The future of materials science lies in integrating the strengths of natural and artificial substances. Bio-inspired materials are being developed that mimic natural structures, such as spider silk or seashells, but are synthesized artificially for industrial use. Nanotechnology enables the design of materials at the atomic level, combining the ecological harmony of natural substances with the versatility of artificial ones. Smart materials that can adapt to environmental changes, self-heal, or respond to stimuli are becoming increasingly important. This fusion of natural inspiration and artificial innovation points toward a future where the distinction between the two may blur, creating a new generation of sustainable and multifunctional materials.

Natural vs. Artificial Materials

1.3.1 Natural Materials

Natural materials are substances obtained directly from nature without significant alteration of their fundamental structure by human processes. They exist in the environment as part of the Earth's natural resources and are classified into categories such as organic, inorganic, metallic, and non-metallic. Examples include wood, stone, clay, cotton, wool, metals such as iron and copper, and minerals like quartz and limestone. These materials are valued for their inherent properties, which evolved naturally over geological or biological timescales. Natural materials played a critical role in the early stages of human civilization, providing tools for survival, building shelters, and forming the basis of trade and culture.

Characteristics of Natural Materials

Natural materials typically exhibit properties that are determined by their origin. Wood, for instance, has a fibrous cellular structure that gives it strength, elasticity, and thermal insulation. Cotton, being plant-derived, has a soft texture and high absorbency, making it ideal for textiles. Stone and clay are naturally durable and weather-resistant, which made them vital for early architecture. Metals extracted from ores often combine strength with ductility, enabling their use in tools, weapons, and construction. One of the defining aspects of natural materials is their variability because no two samples are identical. The texture, color, grain, or purity of natural substances depends on environmental conditions, geography, and the processes that formed them. This variability is both a strength, as it adds uniqueness, and a limitation, as it can restrict uniformity in large-scale applications.

1.3.2 Artificial Materials

Artificial materials, in contrast, are substances created or significantly modified by humans to enhance or alter their properties for specific uses. Unlike natural materials, they do not exist in their final form in nature but are products of human innovation, experimentation, and industrial processes. Artificial materials include plastics, composites, synthetic fibers, ceramics, and alloys that are designed to meet modern demands. For example, nylon and polyester are artificial fibers created to replace or supplement natural textiles like cotton and silk. Plastics such as polyethylene and PVC are engineered to be lightweight, durable, and resistant to corrosion, qualities that natural materials often lack. Artificial materials represent an extension of human capability to manipulate matter, enabling the production of substances tailored to industries such as aerospace, medicine, electronics, and construction.

Characteristics of Artificial Materials

The defining feature of artificial materials is their controllability. Unlike natural materials, they can be produced with uniformity, predictability, and repeatability in large quantities. Engineers can fine-tune their composition, structure, and performance to achieve desired qualities such as heat resistance, tensile strength, flexibility, or transparency. For example, carbon-fiber composites are engineered to combine lightness with immense strength, which is critical in aircraft manufacturing. Artificial materials are also more adaptable than natural ones because they can be designed to perform in extreme environments, such as space, deep oceans, or chemical industries where natural materials would degrade. However, their production often involves energy-intensive processes and chemical treatments that may raise environmental concerns.

1.3.3 Historical Transition from Natural to Artificial

Human history reflects a gradual transition from dependence on natural materials to the creation of artificial substitutes. In prehistoric times, stone, wood, and animal hides formed the basis of survival. The discovery of fire allowed early humans to process natural materials like clay into ceramics and ores into metals, marking the first steps toward artificial transformation.

With the rise of agriculture and settled societies, humans began refining natural fibers such as cotton and wool into textiles. The Industrial Revolution represented a turning point, as new chemical processes led to the creation of artificial dyes, steel, and eventually synthetic polymers. The twentieth century saw an explosion in artificial materials, with plastics and composites dominating industries. Today, most modern technologies, from smartphones to medical implants, rely heavily on artificial materials designed to outperform their natural counterparts in precision and durability.

1.3.4 Comparison of Natural and Artificial Materials

Table: The Contrast between Natural and Man-Made Materials

Aspect	Natural Materials	Artificial (Synthetic) Materials
Origin	Sourced directly from nature (plants, animals, minerals).	Human-made through chemical synthesis or industrial processes.
Examples	Wood, cotton, silk, wool, stone, leather, rubber (natural).	Plastic, nylon, polyester, fiberglass, stainless steel, synthetic rubber.
Composition	Complex, often variable and hete-rogeneous.	Precise, uniform, and controlled.
Production Process	Grown, harvested, or extracted. Requires land, water, and time.	Manufactured in factories through controlled chemical reactions.
Environmental Impact	• Pros: Typically biodegrad-able, renewable (if managed sustainably), low energy pro-duction (often). • Cons: Can require large amou-nts of land/water, agricul-tural chemicals, and processing.	• Pros: Often durable, long-lasting, and can be engineered for specific uses. • Cons: High energy to produce, reliant on fossil fuels, often non-biodegradable, can cause pollution.
Properties	Often unique and variable (e.g., wood grain). Can be biodegrad-able, breathable, and have good aesthetics.	Properties can be precisely engi-neered for a specific purpose (e.g., strength, elasticity, resistance).
Consistency & Quality	Can vary based on source, climate, and season.	Highly consistent and reproducible batch-to-batch.
Cost	Can be inexpensive (e.g., stone) or very expensive (e.g., silk, exotic woods), often subject to market fluctuations.	Often cheaper to mass-produce, with stable pricing. Initial R&D costs can be high.

Key Advantages	Biocompatibility, aesthetic warmth, sustainability (if renewable), traditional appeal.	High performance, durability, functionality, versatility, and availability.
Key Disadvantages	Susceptible to decay, pests, and environmental damage. Supply can be limited.	Pollution from production and disposal, non-renewable feedstocks, can feel "cheap" or artificial.

Natural and artificial materials differ fundamentally in origin, properties, and applications, yet they also share interconnections. Natural materials are limited by availability and inherent characteristics, whereas artificial materials are virtually limitless in design potential. Natural substances often offer better biodegradability and ecological harmony, while artificial materials are valued for efficiency, versatility, and cost-effectiveness in mass production.

Despite their differences, artificial materials frequently mimic or build upon natural ones. For example, synthetic rubber was created to replicate natural rubber but with greater resistance to oil and temperature. Similarly, artificial silk or rayon imitates the luster and feel of natural silk but can be produced more economically. Thus, artificial materials often complement natural ones rather than completely replacing them.

1.3.5 Advantages of Natural Materials

Natural materials have several advantages that have ensured their continued relevance. They are often renewable, such as cotton, wood, or wool, and can be biodegradable, reducing their environmental footprint. Their aesthetic appeal is another strength, as seen in the unique grains of wood, the brilliance of gemstones, or the texture of natural fabrics. Many natural substances also have health and comfort benefits; cotton, for instance, is breathable and hypoallergenic, while clay has cooling properties. Furthermore, natural materials require less processing energy compared to the creation of synthetic alternatives, making them more sustainable in some contexts.

1.3.6 Advantages of Artificial Materials

Artificial materials, however, dominate modern life due to their ability to overcome the limitations of natural ones. They can be designed for specific conditions, such as high temperatures, corrosive environments, or lightweight construction needs. Their uniformity makes them reliable in mass production and ensures consistent performance. Artificial substances are also often more durable, weather-resistant, and versatile in applications ranging from aerospace engineering to household products. In medicine, artificial materials like biocompatible polymers and titanium alloys have made possible implants, prosthetics, and advanced surgical tools. Their capacity to be molded, recycled, or re-engineered further increases their utility across diverse industries.

1.3.7 Environmental Considerations

The environmental impact of natural and artificial materials forms a crucial part of the comparison. Natural materials are generally eco-friendly, especially when sourced responsibly, but overexploitation can lead to deforestation, habitat loss, and resource depletion. Artificial materials, though versatile, often present greater challenges in terms of waste management and pollution. Plastics, for instance, are durable but persist in ecosystems for centuries, creating severe environmental hazards. In response, researchers are developing biodegradable plastics and green composites that combine the advantages of artificial design with environmental sustainability. The balance between natural and artificial materials increasingly depends on finding ways to minimize ecological harm while maximizing performance.

1.4 CLASSIFICATION OF METAMATERIALS

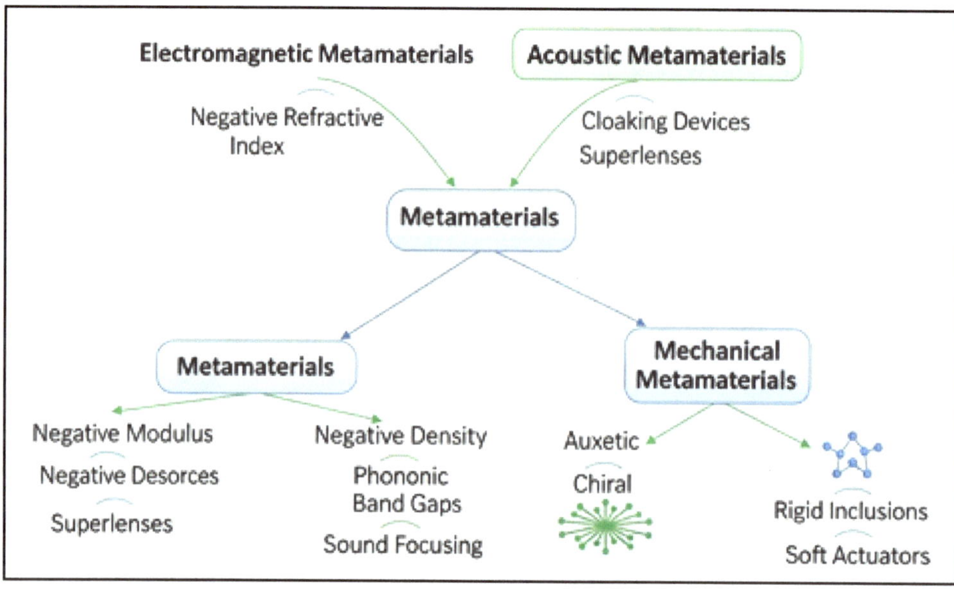

Different Classifications of Metamaterials

The classification of metamaterials provides a systematic way to understand their diversity, properties, and applications. Whether categorized by electromagnetic parameters, frequency regime, dimensional structure, functional application, constituent materials, wave domain, or theoretical framework, each approach highlights a different facet of their remarkable potential. Classifying metamaterials not only organizes knowledge but also guides research toward innovative designs that push the boundaries of what materials can achieve. The multiplicity of classifications also reflects the versatility of metamaterials, which span from electromagnetic to acoustic and thermal domains, from nanoscale optical devices to large-scale seismic shields, and from passive absorbers to adaptive, tunable systems. As material science continues to advance, the classification schemes themselves will evolve, accommodating new discoveries and theoretical breakthroughs. Ultimately, the richness of these classifications underscores the transformative role of metamaterials in modern science and technology, pointing to a future where engineered materials redefine the limits of wave manipulation and human innovation.

Classification by Electromagnetic Properties

One of the most fundamental ways to classify metamaterials is by their electromagnetic properties, particularly their permittivity and permeability. This classification stems from the work of Victor Veselago, who in 1968 theoretically proposed that materials could be engineered to have negative values of both parameters. Based on these characteristics, metamaterials can be grouped into several categories. Single-negative metamaterials have either negative permittivity or negative permeability but not both, leading to unusual propagation of waves with limited applications.

Double-negative metamaterials, also known as left-handed materials, exhibit negative values for both parameters and are capable of producing negative refraction. These materials paved the way for the first experimental verification of Veselago's theory in 2000. There are also epsilon-near-zero metamaterials in which the permittivity approaches zero, enabling unique effects such as tunneling of electromagnetic energy and directive radiation. Similarly, mu-near-zero metamaterials have magnetic permeability close to zero, which allows for unusual guiding of waves and isolation of electromagnetic fields. This electromagnetic classification remains central because it highlights the foundation of metamaterial behavior.

Classification by Frequency Regime

Metamaterials can also be classified according to the frequency range in which they operate. The earliest successful demonstrations were achieved at microwave frequencies because the larger wavelengths allowed researchers to fabricate structures with relative ease. These microwave metamaterials proved useful for radar, antennas, and stealth applications. As nanofabrication techniques advanced, researchers extended metamaterial designs to operate at terahertz and optical frequencies.

Terahertz metamaterials are particularly significant for applications in security scanning, spectroscopy, and wireless communication systems. Optical metamaterials, which function in the visible and near-infrared spectrum, allow control of light in ways that were once thought impossible, leading to advances in imaging, cloaking, and photonic circuits. Acoustic and elastic metamaterials, though not electromagnetic, are also classified by frequency since they control sound and mechanical waves at specific ranges. Thus, classification by frequency regime reflects the technological progress in miniaturization and adaptation of metamaterials to different parts of the electromagnetic spectrum and beyond.

Classification by Dimensional Structure

The geometry and dimensional structure of metamaterials provide another important basis for classification. Three-dimensional bulk metamaterials consist of volumetric structures that extend in all directions, making them suitable for applications like cloaking and superlensing. However, the fabrication of true three-dimensional metamaterials is challenging due to the complexity of manufacturing at small scales. This limitation led to the development of two-dimensional metamaterials, commonly known as metasurfaces.

Metasurfaces are thin, planar structures with engineered patterns that can manipulate the phase, amplitude, and polarization of waves with remarkable precision. Their relative simplicity and efficiency have made them widely studied for practical uses in lenses, holography, and beam steering. In addition to three-dimensional and two-dimensional classifications, one-dimensional metamaterials exist in the form of multilayered stacks, often used to manipulate transmission and reflection of waves across layered boundaries. The classification by dimensionality highlights the different levels of complexity and versatility achievable in metamaterial design.

Classification by Functional Application

Metamaterials are frequently categorized by the functions they perform, which connects their physical properties to practical uses. One major class is absorptive metamaterials, which are designed to absorb electromagnetic radiation almost completely, thereby eliminating reflection and transmission. These are crucial in stealth technology and energy harvesting. Cloaking metamaterials represent another class, using transformation optics to bend waves around an object, rendering it undetectable at certain frequencies.

Superlensing metamaterials belong to a category designed to overcome the diffraction limit of conventional optics, enabling imaging at resolutions below the wavelength of light. Antenna metamaterials improve wireless communication by miniaturizing devices and enhancing bandwidth efficiency. In the acoustic domain, soundproofing metamaterials block or redirect sound waves, while seismic metamaterials are designed to protect structures from earthquakes by redirecting ground vibrations. By classifying metamaterials according to their applications, one can appreciate the breadth of their potential impact across scientific, industrial, and societal fields.

Classification by Constituent Materials

Another basis of classification involves the actual materials used in the construction of metamaterials. Metallic metamaterials were the first to be developed, relying on structures such as split-ring resonators and wire arrays to achieve negative permittivity and permeability. These structures were effective at microwave frequencies but faced significant losses at optical frequencies due to the inherent conductivity of metals.

To overcome these limitations, dielectric metamaterials were introduced, which use non-metallic materials to reduce energy losses while still achieving wave manipulation. Hybrid metamaterials combine metallic and dielectric components to balance the advantages of both. More recently, researchers have developed tunable and reconfigurable metamaterials by incorporating advanced materials such as graphene, liquid crystals, and phase-change substances. These adaptive metamaterials allow dynamic control over their properties in response to external stimuli such as temperature, voltage, or light, opening the door to versatile, real-time applications.

Classification by Wave Domain

Metamaterials can also be classified by the type of wave they interact with, extending beyond the electromagnetic spectrum. Electromagnetic metamaterials manipulate radio, microwave, terahertz, and optical waves. Acoustic metamaterials are designed to control sound waves, achieving phenomena such as acoustic cloaking, sound focusing, and noise cancellation.

Elastic metamaterials extend similar principles to mechanical vibrations, enabling the control of stress and strain in structures. Seismic metamaterials, a subcategory of elastic types, aim to protect buildings from earthquake damage by deflecting or redirecting seismic waves. Thermal metamaterials have also emerged as a distinct class, manipulating heat flow to achieve effects like thermal cloaking or heat concentration. This classification underscores the universality of the metamaterial concept, which is not limited to electromagnetic science but applicable to virtually all wave phenomena.

Classification by Theoretical Framework

A more abstract but important classification system is based on the theoretical framework used to design and describe metamaterials. Effective medium theory classifies metamaterials as homogeneous media with effective permittivity and permeability values, even though they are composed of periodic substructures. Transformation optics provides another framework, where metamaterials are classified by their ability to mimic the bending of space and time on waves, enabling effects such as cloaking and field concentration. Topological metamaterials represent a more recent theoretical classification, where concepts from condensed matter physics are applied to create materials that guide waves along protected edge states, resistant to scattering and defects. This theoretical classification is significant because it emphasizes the underlying physics that governs metamaterial behavior and points toward new frontiers in material science.

Intersections and Hybrid Classifications

While metamaterials can be classified according to properties, frequency, structure, application, materials, domain, or theoretical framework, in practice these categories often intersect. A single metamaterial may belong simultaneously to multiple classifications. For example, a graphene-based metasurface designed for terahertz frequencies could be described by its constituent material, dimensional structure, frequency regime, and functional application. Hybrid metamaterials combine principles across categories, such as thermal-acoustic or electromagnetic-elastic hybrids, to achieve multifunctionality. This intersectional nature of classifications reflects the dynamic and interdisciplinary character of metamaterial research.

Table: Categorizing Metamaterials

Category of Metamaterials	Key Features & Applications
Electromagnetic Metamaterials	Negative refractive index; wave manipulation; used in superlenses, cloaking devices, antennas, radar stealth.
Acoustic Metamaterials	Control sound waves; negative bulk modulus; used in noise cancellation, soundproofing, ultrasound imaging.
Elastic Metamaterials	Control elastic/mechanical waves; negative effective mass density; used for vibration isolation, seismic protection, advanced sensors.
Photonic Metamaterials	Tunable optical bandgaps, slow-light effects; used in optical circuits, laser beam shaping, photonic crystals.
Plasmonic Metamaterials	Surface plasmon resonance, nanoscale light confinement; used for sensing, imaging, energy harvesting, data storage.
Thermal Metamaterials	Heat conduction control, thermal cloaking, anisotropic heat flow; applied in electronics cooling, thermal camouflage.
Mechanical Metamaterials	Auxetic (negative Poisson's ratio), programmable mechanical response; used in medical stents, protective gear, shock absorbers.
Chiral Metamaterials	Optical activity, circular dichroism; applied in polarization control, advanced optical devices.
Topological Metamaterials	Edge states immune to defects, robust wave transport; used in quantum computing, topological insulators, fault-tolerant electronics.

Programmable/Digital Metamaterials	Reconfigurable electromagnetic response, tunable via external control; applied in smart antennas, adaptive optics, IoT systems.
Seismic Metamaterials	Redirect seismic waves, ground vibration control; used in earthquake protection, structural safety systems.
Nonlinear Metamaterials	Intensity-dependent refractive index, harmonic generation; used in nonlinear optics, signal processing, tunable lenses.
Magnetic Metamaterials	Negative permeability, magnetic cloaking; used in MRI enhancement, wireless power transfer.
Hybrid Metamaterials	Combine electromagnetic, acoustic, thermal, or mechanical effects; used in aerospace, multifunctional devices.
Quantum Metamaterials	Quantum coherence effects, nanoscale design; applied in quantum sensing, superconducting devices, quantum communication.

1.4.1 Electromagnetic Metamaterials

Electromagnetic metamaterials embody a revolutionary shift in material science by demonstrating that properties once thought impossible can be achieved through structural design. From Veselago's early theoretical predictions to Pendry's split-ring resonators and Smith's experimental validations, the field has advanced rapidly to encompass superlensing, cloaking, metasurfaces, and tunable devices.

Their classification across frequency domains, structures, and applications reveals the breadth of their influence in science and technology. While challenges remain, particularly in fabrication and energy efficiency, ongoing innovations promise to overcome these barriers. Electromagnetic metamaterials are not merely an academic curiosity but a transformative platform that will continue to reshape optics, communication, and beyond in the twenty-first century.

Despite their promise, electromagnetic metamaterials face challenges. Fabrication at nanoscale precision, especially for optical frequencies, remains complex and costly. Energy losses, particularly in metallic metamaterials, limit efficiency. Scalability for large-area applications also presents obstacles.

Researchers are addressing these issues by developing dielectric metamaterials with lower losses, exploring new fabrication techniques like 3D printing, and pursuing hybrid designs that combine the strengths of different materials. The future of electromagnetic metamaterials lies in integrating them with nanophotonics, quantum technologies, and artificial intelligence for smart, multifunctional devices. As research progresses, they are expected to play a critical role in the advancement of next-generation communication, sensing, and energy systems.

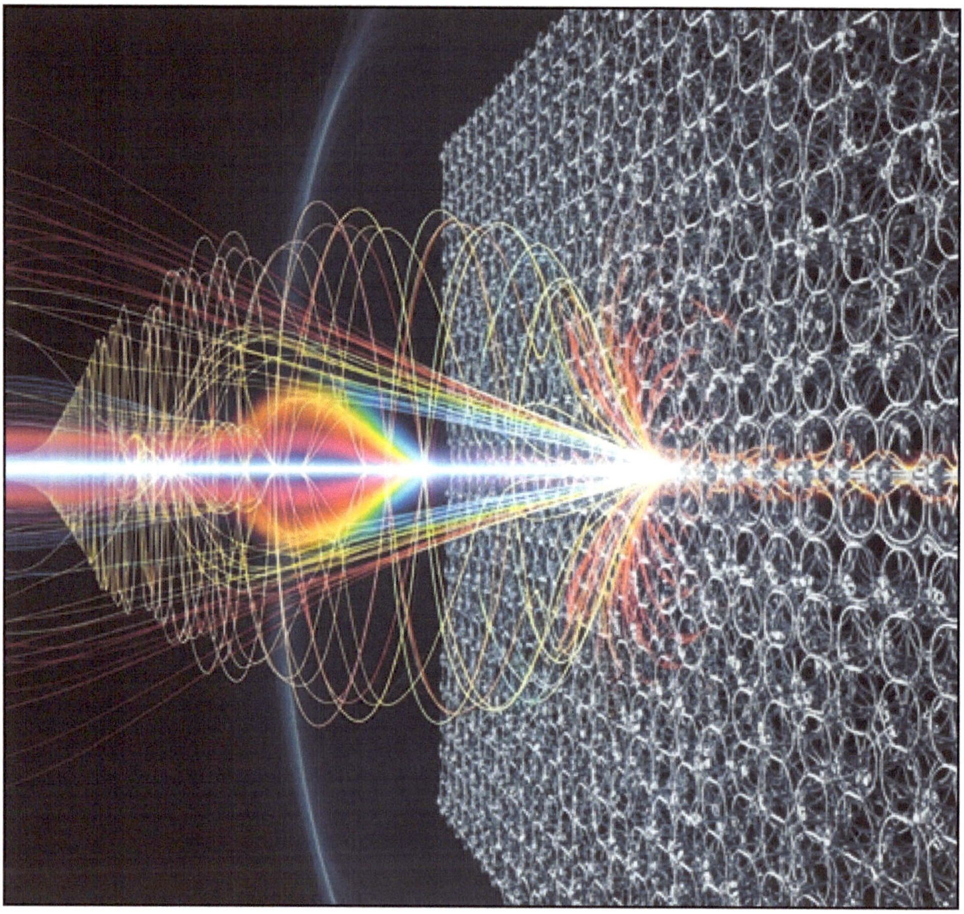

Electromagnetic Metamaterials

Theoretical Foundations

The foundation of electromagnetic metamaterials lies in the work of James Clerk Maxwell, whose equations unified electricity, magnetism, and optics. Maxwell's theory provided the framework for understanding how waves propagate through media characterized by permittivity and permeability.

In 1968, Russian physicist Victor Veselago proposed the possibility of a medium with both negative permittivity and negative permeability, predicting exotic behaviors such as negative refractive index and reversed Doppler and Cherenkov effects. At the time, no natural material exhibited these properties, so Veselago's theory remained speculative for decades. The concept was revived in the late twentieth century when advances in fabrication techniques allowed researchers to design subwavelength structures that could mimic the required electromagnetic responses. This realization marked the birth of electromagnetic metamaterials as an experimental field.

Structural Principles

Electromagnetic metamaterials are constructed from periodic or quasi-periodic arrangements of unit cells, often called meta-atoms, which are significantly smaller than the wavelength of the incident wave. The geometry, orientation, and material composition of these unit cells determine the effective permittivity and permeability of the overall structure. One of the earliest and most influential designs is the split-ring resonator, introduced by Sir John Pendry in the 1990s. The split-ring resonator consists of a pair of concentric metallic rings with gaps, which resonate with magnetic fields and produce negative permeability. When combined with metallic wire arrays that yield negative permittivity, the resulting material can exhibit a negative refractive index. This approach demonstrated that electromagnetic properties could be engineered independently of natural atomic structures, establishing the design principles that underpin modern metamaterials.

Negative Refraction and Left-Handed Materials

One of the defining features of electromagnetic metamaterials is their ability to produce negative refraction. In ordinary right-handed materials, the direction of energy flow and the wave vector are aligned. In contrast, left-handed materials, or double-negative metamaterials, reverse this relationship, leading to counterintuitive behaviors such as waves bending in the opposite direction when entering the medium. This property enables applications like perfect lenses, which can overcome the diffraction limit of conventional optics and achieve super-resolution imaging. The first experimental demonstration of negative refraction occurred in 2000, when David Smith and his team at Duke University constructed a metamaterial composed of split-ring resonators and wire arrays. Their work provided experimental validation of Veselago's predictions and sparked global interest in the practical potential of electromagnetic metamaterials.

Superlensing and Imaging Applications

One of the most promising applications of electromagnetic metamaterials is in superlensing. Traditional lenses are limited by diffraction, which prevents them from resolving features smaller than half the wavelength of light. Pendry proposed that a slab of negative-index metamaterial could amplify evanescent waves, which decay rapidly and are usually lost in conventional imaging.

This would allow the lens to capture fine details beyond the diffraction limit. Experimental efforts have since demonstrated superlensing at microwave, terahertz, and optical frequencies, opening possibilities in fields ranging from biomedical imaging to nanotechnology. The potential to visualize viruses, proteins, and nanoscale structures with unprecedented clarity highlights the transformative role of metamaterials in imaging science.

Cloaking and Transformation Optics

Electromagnetic metamaterials have also enabled the once fantastical idea of cloaking. Using the mathematical framework of transformation optics, researchers can design metamaterials that guide electromagnetic waves around an object, effectively hiding it from detection. The first experimental cloak was demonstrated at microwave frequencies in 2006 by a team at Duke University. Although such cloaks work only for limited frequencies and specific viewing angles, they provided proof of concept for invisibility devices. Beyond cloaking, transformation optics has been used to design concentrators, rotators, and other devices that manipulate electromagnetic fields in novel ways. While practical invisibility cloaks for visible light remain challenging due to fabrication constraints, the theoretical and experimental advances in this area continue to inspire research across optics, defense, and communication technologies.

Frequency Domains of Electromagnetic Metamaterials

Electromagnetic metamaterials are classified according to the frequency regimes in which they operate. Microwave metamaterials were the first to be realized due to the relatively large wavelengths that made fabrication easier. These materials found applications in antennas, radar, and stealth technology. Terahertz metamaterials followed, offering new tools for spectroscopy, imaging, and wireless communication. Extending metamaterial behavior into the visible spectrum required nanofabrication techniques to create structures smaller than hundreds of nanometers. Optical metamaterials have since enabled unprecedented control over visible light, with applications in compact photonic devices, holography, and optical data storage. Across each frequency domain, electromagnetic metamaterials expand the ability to manipulate waves and create devices tailored to specific technological needs.

Metasurfaces and Two-Dimensional Designs

While three-dimensional metamaterials demonstrate remarkable properties, they often present fabrication challenges. This led to the rise of metasurfaces, which are two-dimensional counterparts that manipulate electromagnetic waves through subwavelength patterns on thin surfaces. Metasurfaces can control amplitude, phase, and polarization with high precision, enabling flat lenses, holograms, and beam-steering devices. Their compactness and efficiency make them more practical for integration into devices like cameras, displays, and sensors. Metasurfaces represent a major evolution in the field, making electromagnetic metamaterial principles more accessible to real-world technologies.

Tunable and Reconfigurable Electromagnetic Metamaterials

A significant advancement in the field is the development of tunable and reconfigurable electromagnetic metamaterials. Traditional designs are static, meaning their properties are fixed once fabricated. To address this limitation, researchers have introduced materials like graphene, liquid crystals, and phase-change substances that allow real-time control over electromagnetic responses. For example, applying an electric voltage to a tunable metamaterial can shift its resonant frequency, making it adaptable to changing environments. These reconfigurable designs are crucial for next-generation communication systems, adaptive imaging, and dynamic cloaking devices. The ability to actively control electromagnetic behavior reflects the growing sophistication of metamaterial research.

Applications in Communication and Technology

Electromagnetic metamaterials have already made significant contributions to communication and technology. In antenna engineering, they improve performance by reducing size while enhancing bandwidth and gain. This is particularly valuable in mobile devices, satellite communication, and emerging 6G wireless networks.

In defense applications, metamaterials enhance stealth technology by absorbing radar waves or guiding them away from detection. In energy harvesting, they enable efficient solar concentrators and wireless power transfer systems. In medicine, metamaterial-based imaging devices improve diagnostics, while novel sensors detect subtle changes in electromagnetic fields. These diverse applications illustrate the transformative potential of electromagnetic metamaterials across multiple industries.

Table: Electromagnetic Metamaterials Based on their Properties, Mechanisms and Applications

Property / Mechanism	Applications
Negative Refractive Index (NRI)	Superlenses, invisibility cloaks, subwavelength imaging
Negative Permittivity ($\varepsilon < 0$)	Waveguides, shielding, focusing beyond diffraction limit
Negative Permeability ($\mu < 0$)	Magnetic cloaking, compact antennas, microwave devices
Double-Negative ($\varepsilon < 0, \mu < 0$)	Perfect lensing, radar evasion, advanced communication systems
Anisotropic Response	Polarization control, optical filters, telecom wave manipulation
Dispersion Engineering	Slow-light devices, enhanced data transmission
Hyperbolic Dispersion	Nanophotonics, near-field imaging, high-density optical circuits

Near-Zero Index (ENZ, MNZ, DNZ)	Wave tunneling, directive antennas, impedance matching
Metamaterial Absorbers	Stealth coatings, photovoltaic enhancement, sensors
Nonlinear Effects	Optical switches, frequency mixers, tunable photonic devices
Chiral/Optically Active Structures	Polarization converters, circular polarizers, optical isolators
Reconfigurable/Tunable Metamaterials	Smart antennas, adaptive optics, wireless communication systems
Plasmonic Resonance	Biosensing, chemical detection, surface-enhanced spectroscopy
Gradient-Index (GRIN) Metamaterials	Beam shaping, flat lenses, cloaking
Topological EM States	Robust photonic circuits, fault-tolerant communication links

1.4.2 Acoustic Metamaterials

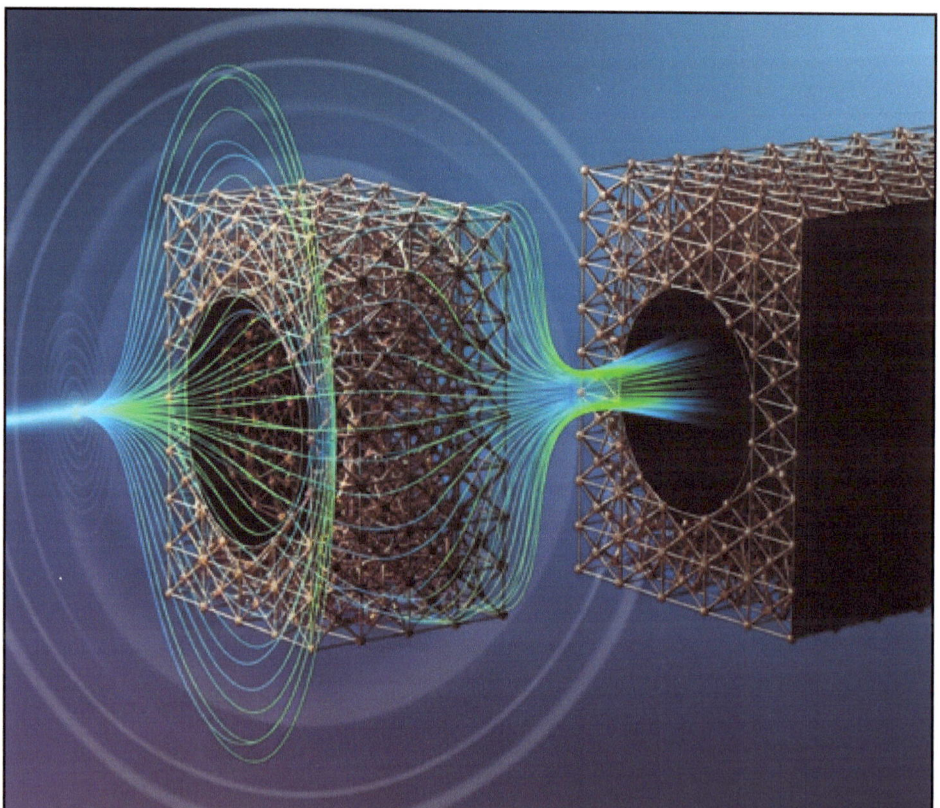

Acoustic Metamaterials

Acoustic metamaterials represent a revolutionary step forward in the science of sound manipulation. By harnessing engineered structures rather than natural properties, they enable unprecedented control over acoustic waves, from cloaking and superlensing to soundproofing and medical imaging. Their classification across structural principles, frequency domains, and applications demonstrates their versatility and wide-ranging potential.

Although challenges remain in fabrication and scalability, ongoing research is rapidly pushing the boundaries of what these materials can achieve. As acoustic metamaterials continue to mature, they promise to transform noise management, healthcare, communication, and industrial technologies, marking a new era where the manipulation of sound is limited only by human imagination and innovation.

Despite their progress, acoustic metamaterials face challenges in scalability, fabrication, and broadband operation. Many designs work only within narrow frequency ranges, limiting their practical use. Manufacturing complex resonator structures at large scales can also be costly and time-consuming. Researchers are addressing these issues through advanced fabrication techniques such as 3D printing and by developing hybrid metamaterials that combine multiple mechanisms to broaden frequency responses. The future of acoustic metamaterials also lies in interdisciplinary collaboration, where insights from physics, materials science, and engineering converge to create multifunctional and sustainable solutions. As the field evolves, acoustic metamaterials are expected to play a vital role in shaping the way humans interact with sound in daily life, medicine, and industry.

Origins and Conceptual Foundations

The concept of acoustic metamaterials emerged in the early twenty-first century, inspired by advances in electromagnetic metamaterials. Researchers realized that many of the theoretical principles used to manipulate electromagnetic waves could also be applied to acoustic waves, which are mechanical vibrations that propagate through a medium such as air, water, or solids. In 2000, the first experimental demonstrations of electromagnetic metamaterials sparked curiosity about extending these ideas to sound. By 2009, researchers had successfully created acoustic metamaterials with negative effective properties, proving that sound waves could be engineered in ways previously considered impossible. These discoveries established a new field of research where the emphasis shifted from relying on natural sound propagation to designing artificial structures for unprecedented control.

Structural Principles of Acoustic Metamaterials

The extraordinary behavior of acoustic metamaterials arises from their engineered unit cells, often called meta-atoms or resonators, which are much smaller than the wavelength of sound they manipulate. These unit cells can take various forms such as Helmholtz resonators, membranes, or coiled channels. Helmholtz resonators mimic the acoustic properties of bottles or cavities with narrow necks, enabling control over resonance frequencies. Membrane-based metamaterials rely on thin elastic layers with attached masses, which can trap and manipulate low-frequency sound waves. Coiled space structures, sometimes referred to as "acoustic labyrinths," extend the effective path of sound within a compact volume, enabling materials to achieve properties not possible in natural substances. The design flexibility of these unit cells allows researchers to tailor acoustic responses across a wide frequency range.

Negative Parameters and Acoustic Wave Control

One of the defining features of acoustic metamaterials is the ability to achieve negative effective parameters, analogous to negative permittivity and permeability in electromagnetic metamaterials. In acoustics, the key parameters are effective mass density and bulk modulus. Negative mass density occurs when the acceleration of the medium is opposite to the applied force, creating counterintuitive wave propagation. Negative bulk modulus arises when pressure increases result in expansion rather than compression. Materials that exhibit both negative mass density and negative bulk modulus are known as double-negative acoustic metamaterials, which enable negative refraction of sound waves. These unusual properties provide the foundation for applications such as superlensing and cloaking in the acoustic domain.

Acoustic Cloaking and Invisibility

One of the most intriguing applications of acoustic metamaterials is acoustic cloaking, where sound waves are guided around an object so that it becomes undetectable to acoustic sensors. Using the principles of transformation acoustics, researchers have designed materials that bend sound paths smoothly around obstacles without scattering. Experimental demonstrations have shown partial cloaking for ultrasonic waves, where objects are rendered acoustically invisible in certain frequency ranges. Although practical full-spectrum acoustic cloaks remain a challenge, the progress made so far highlights the potential for stealth technologies, noise reduction in architectural spaces, and even medical imaging techniques where cloaking could isolate specific regions from sound waves.

Superlensing and Subwavelength Imaging

Another breakthrough enabled by acoustic metamaterials is the concept of the acoustic superlens. Conventional acoustic lenses, like their optical counterparts, are limited by diffraction and cannot resolve features smaller than half the wavelength of sound. Acoustic metamaterials with negative refraction overcome this limitation by amplifying evanescent waves, allowing subwavelength imaging. This has profound implications in fields such as medical ultrasound, where super-resolution imaging could allow doctors to observe tissues and cells at scales beyond the reach of conventional devices. Non-destructive testing and materials evaluation also benefit from acoustic superlensing, enabling engineers to detect micro-cracks or defects with enhanced precision.

Frequency Domains of Acoustic Metamaterials

Acoustic metamaterials are designed to operate in different frequency regimes depending on the application. At low frequencies, they are especially valuable because natural materials are often ineffective at controlling long-wavelength sound waves. Membrane-based and labyrinthine designs excel in this regime, making them useful for noise cancellation in buildings, vehicles, and industrial machinery. At ultrasonic frequencies, acoustic metamaterials find applications in medical imaging and high-resolution sensing. Their ability to manipulate sound across wide frequency ranges demonstrates their versatility and adaptability to different technological needs.

Metasurfaces in Acoustic Applications

Similar to electromagnetic metasurfaces, acoustic metasurfaces have emerged as thin, planar structures capable of manipulating sound with high precision. Instead of relying on bulky three-dimensional metamaterials, metasurfaces achieve phase, amplitude, and wavefront control using compact, engineered patterns.

Acoustic metasurfaces have been used to create flat acoustic lenses, sound holograms, and beam shapers. Their lightweight and easy-to-integrate designs make them particularly attractive for architectural acoustics, personal audio devices, and portable medical technologies. By reducing size and complexity, acoustic metasurfaces make metamaterial concepts more practical for real-world applications.

Tunable and Active Acoustic Metamaterials

Traditional acoustic metamaterials are passive, meaning their properties remain fixed once fabricated. However, recent advances have introduced tunable and active acoustic metamaterials that adapt dynamically to changing environments. These designs incorporate elements such as piezoelectric actuators, fluids, or smart materials that respond to external stimuli.

By adjusting voltage, pressure, or temperature, researchers can alter the acoustic response of the material in real time. This adaptability opens possibilities for smart noise-canceling walls, reconfigurable acoustic imaging systems, and responsive medical devices. Active metamaterials represent the next stage in the evolution of acoustic wave control, merging engineering precision with environmental responsiveness.

Applications in Noise Reduction and Soundproofing

One of the most practical and immediately beneficial applications of acoustic metamaterials is in noise reduction and soundproofing. Traditional materials like foams and panels are often ineffective against low-frequency sounds, which are difficult to absorb or block due to their long wavelengths. Acoustic metamaterials overcome this limitation by using resonant structures to trap and dissipate low-frequency energy efficiently. This makes them highly desirable for reducing noise pollution in urban environments, improving acoustic comfort in buildings, and minimizing noise in vehicles. Industries such as aviation and automotive engineering are particularly interested in these technologies to enhance passenger comfort and reduce environmental noise impact.

Medical and Industrial Applications

Beyond noise reduction, acoustic metamaterials have promising applications in medicine and industry. In healthcare, metamaterials are improving ultrasound imaging and therapy. Super-resolution imaging allows for earlier and more accurate detection of diseases, while focused ultrasound therapies benefit from precise control of sound waves within the body. In industrial settings, acoustic metamaterials are being developed for non-destructive testing, allowing engineers to inspect structures for flaws without damaging them. Additionally, they hold promise in sonar technologies for underwater exploration and defense. The breadth of these applications underscores the transformative potential of acoustic metamaterials in multiple sectors.

1.4.3 Mechanical Metamaterials

Mechanical metamaterials represent a revolution in the way we understand and design materials. By shifting focus from chemical composition to geometric architecture, they enable extraordinary mechanical behaviors such as negative Poisson's ratio, negative compressibility, and programmable elasticity.

Their applications span a wide range of fields, from aerospace and robotics to medicine and protective systems. Although challenges remain in scalability and durability, ongoing advances in design and fabrication promise to overcome these hurdles. As research progresses, mechanical metamaterials are expected to play a pivotal role in shaping the future of engineering and technology, where imagination and structural creativity define the limits of material performance. Despite their promise, mechanical metamaterials face several challenges. Scalability remains a major issue, as many designs are demonstrated at laboratory scales but are difficult to produce cost-effectively on an industrial scale. Durability and long-term performance in real-world conditions are also concerns, particularly for applications in aerospace and biomedical fields. Furthermore, balancing exotic properties with practical usability requires careful design to ensure that metamaterials meet safety, environmental, and economic requirements. Looking ahead, interdisciplinary collaboration between materials scientists, engineers, physicists, and biologists will be essential to unlock the full potential of mechanical metamaterials. The future likely holds materials that are not only mechanically exceptional but also multifunctional, combining structural, thermal, and acoustic properties into integrated systems.

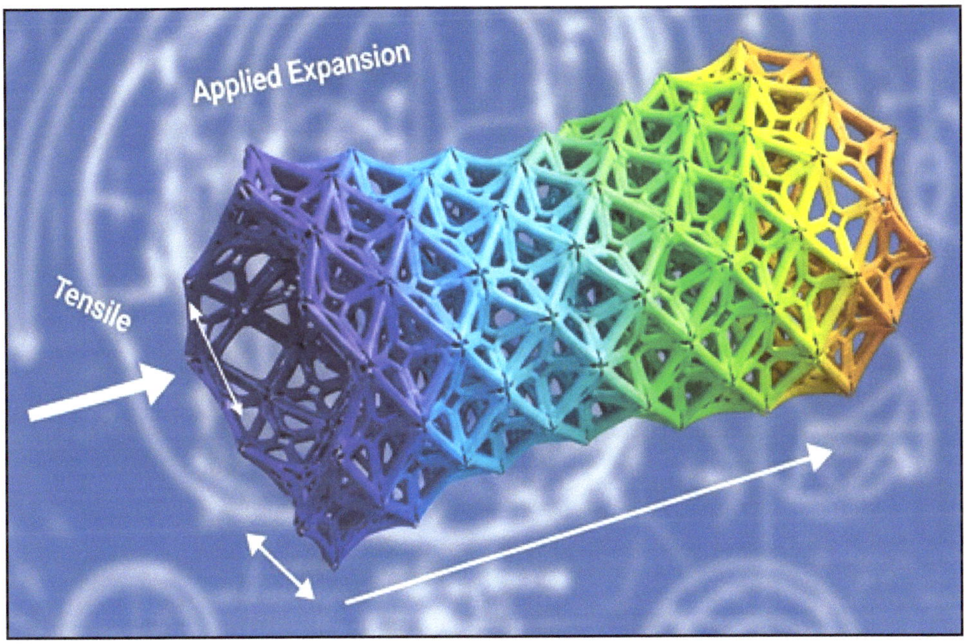

Mechanical Metamaterials

Concept and Theoretical Foundations

The foundation of mechanical metamaterials lies in the principle that geometry governs mechanical behavior as much as, if not more than, material composition. For centuries, engineers and scientists viewed strength, stiffness, and ductility as properties determined largely by a substance's internal chemistry. However, the advent of advanced design tools, computational modeling, and additive manufacturing techniques has demonstrated that mechanical performance can be controlled by structuring materials into periodic or patterned architectures. By embedding specific patterns into materials at scales smaller than the structural response length, researchers have engineered systems where stress, strain, and deformation follow pathways that do not exist in natural materials. This structural determinism has enabled the development of lightweight yet ultra-strong designs and materials with properties that defy conventional intuition.

Negative Poisson's Ratio and Auxetic Materials

One of the most well-known breakthroughs in mechanical metamaterials is the creation of auxetic materials, which exhibit a negative Poisson's ratio. In most natural materials, stretching in one direction causes contraction in the perpendicular direction. Auxetic materials behave in the opposite way; when stretched, they expand laterally, and when compressed, they contract in all directions. This unique property results from specially designed geometries such as re-entrant honeycombs, rotating squares, and chiral structures. Auxetic materials offer significant advantages, including enhanced energy absorption, improved shear resistance, and superior indentation resistance. Applications of auxetic metamaterials are being explored in medical implants, sports equipment, body armor, and vibration damping systems, where their unique expansion behavior provides increased performance and safety.

Negative Compressibility and Unusual Elasticity

Another striking phenomenon achievable through mechanical metamaterials is negative compressibility, where materials contract under tension and expand under compression, defying conventional mechanical logic. Such behavior is achieved through carefully designed bistable or multistable structures that switch between stable states under stress. These unusual elastic responses open opportunities for mechanical switches, actuators, and shock-absorbing systems. Mechanical metamaterials can also demonstrate extreme tunability in elasticity, allowing stiffness to be adjusted across wide ranges. Programmable elasticity is particularly attractive in robotics, where adaptable materials can provide both rigidity and flexibility depending on the operational need.

Lightweight and High-Strength Architectures

Mechanical metamaterials are not only about exotic properties but also about achieving high performance in practical applications. One of the most important directions in research has been the development of lightweight yet high-strength architectures. By employing geometries such as lattice frameworks, trusses, and cellular structures, scientists have engineered materials that achieve high strength-to-weight ratios. These designs mimic natural inspirations like bone, wood, and shells but are optimized through mathematical and computational modeling. In aerospace engineering, where weight reduction directly translates into fuel efficiency, mechanical metamaterials offer a path to stronger yet lighter aircraft and spacecraft components.

Programmable and Reconfigurable Mechanical Metamaterials

A rapidly growing area within the field is programmable and reconfigurable mechanical metamaterials. Unlike conventional materials with fixed properties, these structures can be tuned dynamically to alter their mechanical response. Through designs that incorporate hinges, folds, or embedded responsive materials, metamaterials can change stiffness, shape, or damping characteristics in real time. Origami-inspired and kirigami-inspired structures have been particularly influential in this domain. These approaches use folding and cutting patterns to create materials that are compact, deployable, and reconfigurable. Such designs have applications in space exploration, where foldable structures can be compactly stored during launch and deployed to large functional systems in orbit.

Topological Mechanical Metamaterials

Topological concepts, originally introduced in physics and mathematics, have recently been applied to mechanical metamaterials to create systems with robust wave propagation properties. Topological mechanical metamaterials are designed so that vibrations or mechanical waves travel along their edges or interfaces without being scattered by defects or disorder. This phenomenon ensures highly reliable and predictable mechanical wave transport, which is valuable for vibration control, signal transmission, and structural health monitoring. The robustness of these topological states is rooted in geometry and symmetry, providing new strategies for designing materials that can guide mechanical energy with minimal loss.

Acoustic and Vibration Control

Mechanical metamaterials also overlap with acoustic metamaterials in the control of vibrations and sound. Structures with resonant units or periodic arrangements can filter, block, or guide vibrations in desired ways. These properties are especially useful in engineering structures exposed to continuous vibrations, such as bridges, buildings, and vehicles. By embedding mechanical metamaterials into these structures, it becomes possible to isolate vibrations, dampen seismic waves, and enhance overall stability. This cross-disciplinary aspect highlights how mechanical metamaterials can be multifunctional, serving both structural and acoustic purposes simultaneously.

Biomedical Applications of Mechanical Metamaterials

The biomedical field has embraced mechanical metamaterials due to their adaptability, biocompatibility, and unique mechanical performance. Auxetic implants, for example, are being developed to mimic the mechanical behavior of human tissues, allowing for better integration with bone and cartilage. Metamaterial scaffolds with tunable porosity and stiffness are being explored in tissue engineering, enabling customized support for cell growth and regeneration. In prosthetics, lightweight and resilient mechanical metamaterials improve both comfort and durability. Additionally, protective gear such as helmets and padding made with mechanical metamaterials provides superior impact absorption, ensuring enhanced safety for athletes and soldiers.

Energy Absorption and Protective Systems

Mechanical metamaterials are highly effective in energy absorption, making them ideal for protective applications. Structures designed to collapse in controlled ways under impact can absorb large amounts of energy while minimizing damage to the protected object or person. Auxetic foams and lattice-based metamaterials are being used to create next-generation helmets, car bumpers, and crash-protection systems. Their ability to distribute impact forces uniformly and resist penetration far surpasses traditional protective materials. This makes them indispensable in industries where safety and reliability are paramount.

Fabrication Techniques and Design Advances

The development of mechanical metamaterials has been accelerated by advancements in fabrication techniques such as 3D printing, microfabrication, and additive manufacturing. These technologies allow the creation of intricate and highly precise structures that were previously impossible to build. Computational modeling and optimization algorithms have also played a crucial role, enabling researchers to test thousands of design variations virtually before physical production. The synergy between digital design tools and advanced fabrication methods ensures that mechanical metamaterials continue to evolve rapidly, with increasingly complex and multifunctional architectures.

1.4.4 Thermal Metamaterials

Unique Properties / Mechanisms and their Applications

- **Auxetic Behavior (Negative Poisson's Ratio):** Expand laterally when stretched—Medical stents, protective padding, flexible armor.
- **Ultra-Lightweight Lattices:** 3D micro-architectures with very low density—Aerospace structures, lightweight vehicles, energy-efficient design.
- **High Energy Absorption:** Ability to dissipate large amounts of energy—Crash protection in vehicles, helmets, impact-resistant gear.
- **Programmable Mechanical Response:** Tunable stiffness, strength, or shape—Soft robotics, adaptive materials, deployable structures.
- **Shape Memory Effect:** Regain original shape upon heating or stimulus—Actuators, biomedical implants, morphing aerospace surfaces.
- **Mechanical Cloaking:** Redirect mechanical stress around objects—Protecting sensitive components, structural safety.

- **Fracture Resistance:** Micro-architectures that prevent crack propagation—Long-lasting medical implants, durable aerospace materials.
- **Vibration Damping:** Lattice structures tuned for mechanical wave absorption—Seismic protection, vibration isolation in machinery.
- **Buckling-Induced Functionality:** Controlled deformation under stress—Energy absorbers, tunable mechanical devices.
- **Topological Mechanical States:** Edge modes resistant to defects—Robust signal transmission in mechanical systems, vibration control.
- **Hierarchical Structures:** Multiple scales of design for strength and lightness—Bone-mimicking implants, aerospace panels.
- **Bio-Inspired Designs:** Mimic natural structures like bone, shells, honeycombs—Prosthetics, strong lightweight packaging, biomimetic devices.
- **Tunable Elasticity:** Adjustable stiffness depending on load/stimulus—Artificial muscles, adaptive structural components.
- **Soft Mechanical Metamaterials:** Rubber-like lattices with programmable response—Soft robotics, wearable devices, medical devices.
- **Origami/Kirigami-Based Structures:** Folding/cutting patterns for reconfigurability—Deployable shelters, and space structures.
- **Negative Compressibility:** Contract when tensioned and expand under compression—Advanced actuators, adaptive shock absorbers.
- **Acoustic Mechanical Coupling:** Mechanical lattices interacting with sound waves—Noise control, sound insulation, acoustic imaging.
- **Thermo-Mechanical Coupling:** Respond mechanically to thermal stimuli—Heat-activated actuators, smart textiles.
- **Magneto-Mechanical Coupling:** Mechanical behavior controlled by magnetic fields—Sensors, actuators, tunable damping devices.
- **Electro-Mechanical Coupling:** Integrate electrical response with mechanical deformation—Energy harvesters, responsive sensors, flexible electronics.
- **Multi-Stability:** Existence of multiple stable configurations—Switchable structures, memory devices, reconfigurable robotics.
- **3D Printing/Advanced Fabrication Enabled Designs:** Micro-architected via additive manufacturing—Custom medical implants, aerospace components, rapid prototyping.

- **Porous Mechanical Metamaterials:** Engineered porosity for lightness and fluid interaction—Tissue scaffolds, filtration systems, lightweight structures.
- **Gradient Mechanical Properties:** Varying stiffness/strength across material Protective gear, adaptive prosthetics, impact mitigation.
- **Seismic Mechanical Metamaterials:** Redirection of ground vibrations—Earthquake protection for buildings and bridges.

Concept and Foundations

The fundamental concept of thermal metamaterials lies in controlling the conduction and diffusion of heat through engineered structures. Traditional thermodynamics states that heat naturally flows from regions of high temperature to regions of low temperature, following Fourier's law of heat conduction. Natural materials can accelerate or slow this process depending on their thermal conductivity, but they cannot fundamentally alter the pathways of heat flow in extraordinary ways.

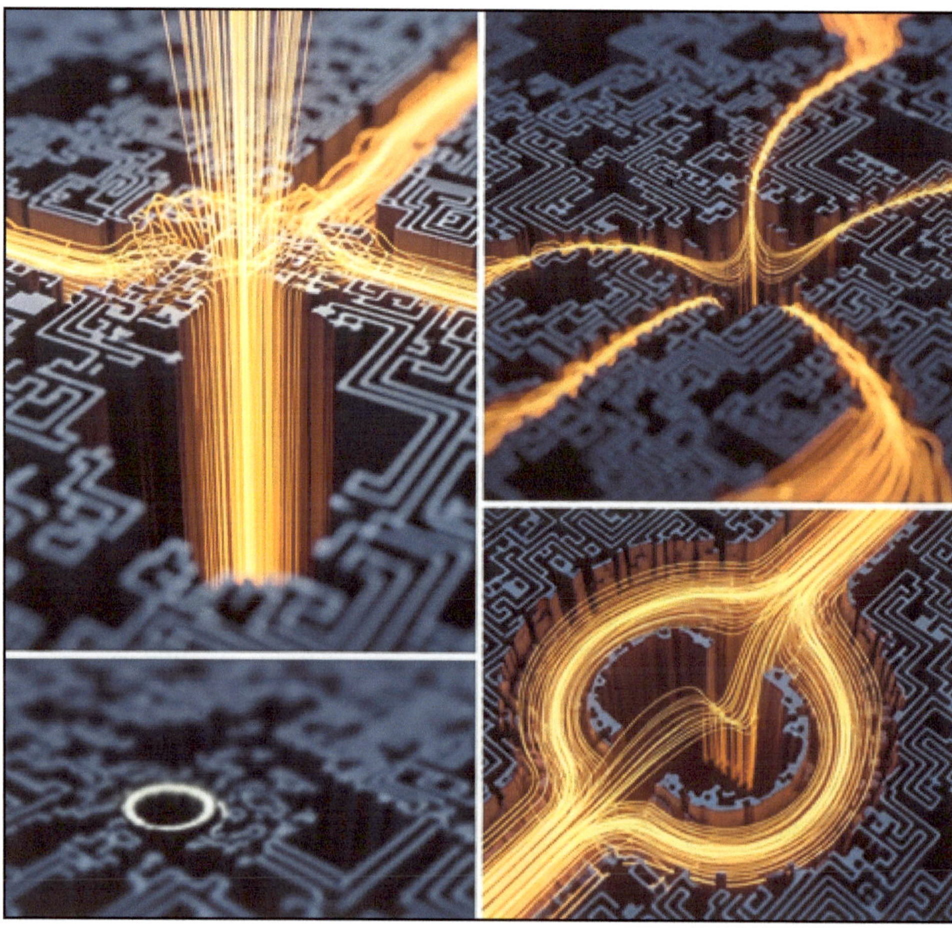

Thermal Metamaterials

Thermal metamaterials challenge this limitation by structuring materials so that heat can be guided, bent, or even cloaked around objects. These effects are achieved through principles of transformation thermotics, which adapt the mathematics of coordinate transformations originally applied to electromagnetic metamaterials into the realm of thermal physics. By carefully tailoring the geometry and arrangement of substructures, thermal metamaterials create effective thermal properties that go beyond natural possibilities.

Transformation Thermotics and Heat Control

Transformation thermotics forms the theoretical backbone of thermal metamaterials. Just as transformation optics allows the bending of light by distorting spatial coordinates, transformation thermotics manipulates the heat equation to control the flow of thermal energy.

Through this method, researchers can design materials where heat flows around an object without disturbing its surroundings, effectively rendering it thermally invisible. Similarly, transformation thermotics allows the concentration of heat into a small region, amplifying thermal effects for energy harvesting or thermal sensing. This theoretical framework provides the mathematical foundation that makes advanced functions like cloaking and concentration possible in practice.

Thermal Cloaking and Invisibility

One of the most remarkable breakthroughs in thermal metamaterials is the demonstration of thermal cloaking. A thermal cloak hides an object from thermal detection by guiding heat around it in such a way that the temperature field outside remains undisturbed. This means that infrared sensors or thermal imaging devices cannot detect the object, as it leaves no trace in the temperature distribution. Experimental demonstrations have shown successful cloaking at room temperature using composite materials with graded thermal conductivities. While perfect cloaking across all conditions remains a challenge, thermal cloaks already hold significant promise in defense, privacy, and industrial applications where thermal signatures must be controlled or concealed.

Heat Concentrators and Thermal Lenses

Another significant application of thermal metamaterials is the creation of heat concentrators or thermal lenses. These devices collect heat from a large area and focus it into a smaller region, much like an optical lens focuses light. This ability to concentrate heat is particularly useful in energy harvesting technologies, where maximizing thermal gradients improves efficiency.

For example, thermoelectric generators can benefit from thermal concentrators that enhance the conversion of waste heat into usable electricity. Similarly, thermal lenses can be applied in localized heating for medical therapies, where precise temperature control is necessary to target tumors or damaged tissues without affecting surrounding areas.

Thermal Diodes and Rectifiers

Thermal metamaterials also pave the way for the development of thermal diodes and rectifiers, devices that control the directionality of heat flow. In natural systems, heat transfer is generally reciprocal, meaning that heat flows equally well in both directions between two regions. Thermal diodes break this symmetry by allowing preferential heat conduction in one direction while suppressing it in the opposite. This is achieved through asymmetric material design and nonlinear thermal responses. Thermal diodes and rectifiers are vital for thermal management in electronic circuits, where preventing backflow of heat can protect sensitive components and improve overall performance. They also represent an important step toward the development of thermal logic circuits, where heat rather than electricity serves as the carrier of information.

Thermal Camouflage and Stealth Applications

Beyond cloaking and directionality, thermal metamaterials also enable thermal camouflage, where an object is made to appear as though it has the same temperature as its background. Unlike cloaking, which completely hides the object from heat flow, camouflage manipulates the thermal signature so that detection devices cannot distinguish the object from its environment. Thermal camouflage is especially relevant for military stealth applications, where vehicles, equipment, or personnel must avoid detection by infrared imaging systems. Additionally, it can be applied in civilian contexts such as privacy technologies, wildlife research, and even architectural design for energy efficiency by masking heat loss in buildings.

Active and Tunable Thermal Metamaterials

While many early thermal metamaterials were passive, meaning their properties were fixed once designed, recent advances have introduced active and tunable systems. These materials incorporate smart components that respond to external stimuli such as electric fields, magnetic fields, or mechanical stress, allowing real-time control of thermal properties. For example, thermally tunable metamaterials can switch between cloaking and concentrating modes depending on operational requirements. Phase-change materials such as vanadium dioxide are also integrated into metamaterial structures to dynamically alter thermal conductivity as temperatures change. These innovations enable adaptive thermal management systems that can adjust to varying environments, making them useful in space exploration, smart electronics, and next-generation climate control systems.

Applications in Electronics Cooling

One of the most immediate and practical applications of thermal metamaterials lies in electronics cooling. As electronic devices become smaller and more powerful, heat dissipation has become a critical challenge. Overheating not only reduces performance but also shortens the lifespan of components. Thermal metamaterials provide innovative solutions by guiding heat away from hotspots, spreading it evenly across surfaces, or concentrating it into designated heat sinks. This precise thermal management is essential for high-performance computing, mobile devices, and next-generation processors, ensuring efficiency and reliability.

Energy Harvesting and Sustainability

Thermal metamaterials also play an important role in energy harvesting and sustainability. Waste heat is a major byproduct of industrial processes, transportation systems, and power plants, often dissipated unused into the environment. By employing thermal metamaterials, waste heat can be concentrated and directed toward thermoelectric devices for efficient conversion into electricity. This contributes to energy conservation and reduces environmental impact. Furthermore, thermal metamaterials can improve solar thermal systems by focusing sunlight-induced heat into specific regions, enhancing efficiency in renewable energy technologies.

1.5 FUNDAMENTAL PROPERTIES AND BEHAVIORS

The fundamental properties and behaviors of metamaterials represent a profound shift in how scientists and engineers understand and manipulate waves. From negative refractive index and negative parameters to cloaking, bandgap formation, and nonlinear tunability, metamaterials offer a toolkit of extraordinary capabilities that defy conventional expectations. Their unique behaviors arise not from chemistry but from architecture, where the careful design of unit cells gives rise to macroscopic effects unseen in nature. Although challenges remain in addressing losses, dispersion, and scalability, the progress made in this field underscores its transformative potential. As research continues, the fundamental properties and behaviors of metamaterials will not only inspire scientific curiosity but also drive innovations in optics, acoustics, thermal management, telecommunications, and medical technologies, marking a new frontier in the control of waves.

1.5.1 Negative Refractive Index

One of the most defining and revolutionary properties of metamaterials is their ability to exhibit a negative refractive index. In ordinary materials, when light passes from one medium to another, it bends according to Snell's law, with the angle determined by the positive refractive indices of the two media. Metamaterials can bend light in the opposite direction, creating what is known as negative refraction. This phenomenon occurs because metamaterials can be engineered to simultaneously achieve negative values of effective permittivity and permeability. Negative refraction opens the door to unusual optical phenomena such as superlensing, where imaging can surpass the diffraction limit, and cloaking, where light waves can be guided around objects without detection.

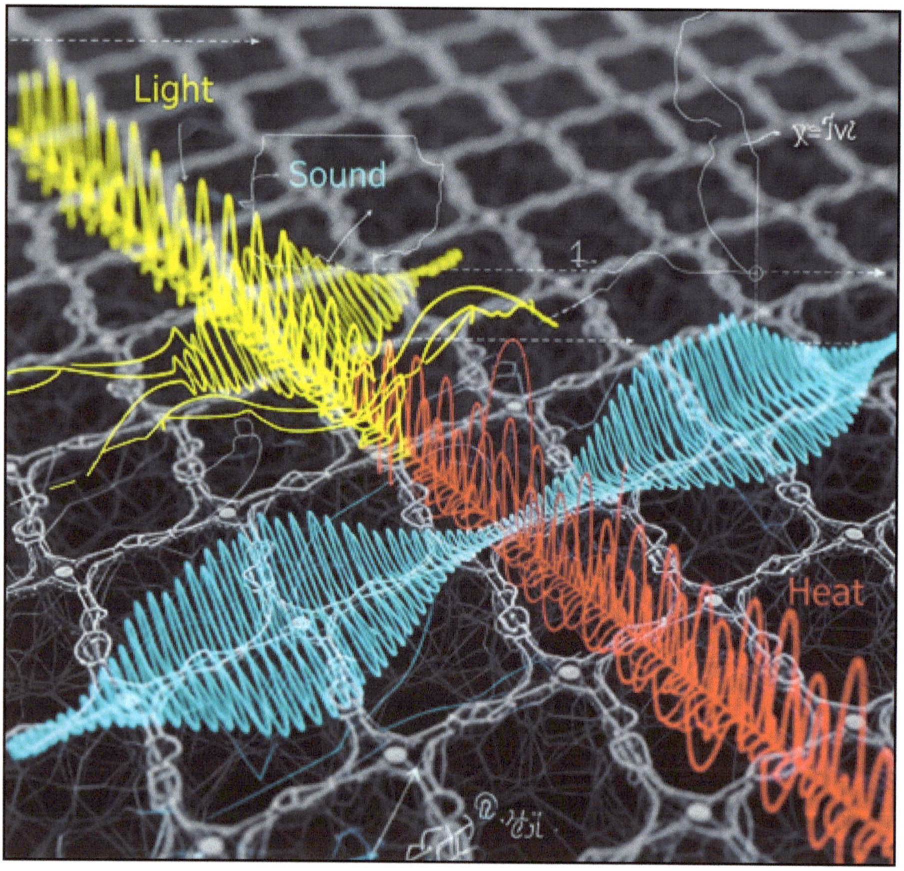

The Governing Characteristics and Macroscopic Behavior of Metamaterials

1.5.2 Effective Medium Approximation

The behaviors of metamaterials are often described using the concept of effective medium approximation. In this framework, the metamaterial is treated as a homogeneous medium even though it is composed of periodic or patterned structures on a microscopic scale. The unit cells, which are smaller than the relevant wavelength, collectively produce averaged properties such as effective permittivity, permeability, density, or modulus. These effective parameters allow metamaterials to be characterized in ways similar to natural materials, but with engineered values that may be positive, negative, or near zero. The effective medium approximation is central to understanding how local geometry can dictate global wave behavior.

1.5.3 Anisotropy and Directional Dependence

Many metamaterials are designed to be anisotropic, meaning their properties vary with direction. Unlike isotropic natural materials, which behave uniformly in all directions, anisotropic metamaterials can be engineered to control wave propagation differently along different axes.

This directional dependence allows for precise manipulation of wave paths, enabling applications such as beam steering, polarization control, and acoustic focusing. Anisotropy also plays a vital role in cloaking devices, where waves must be carefully guided around objects without reflection or scattering. By controlling anisotropic responses, researchers achieve a higher degree of control over how waves travel through space.

1.5.4 Bandgap Formation and Wave Filtering

Another fundamental behavior of metamaterials is the creation of bandgaps, frequency ranges in which waves cannot propagate. Similar to electronic bandgaps in semiconductors, metamaterial bandgaps arise from periodic structural arrangements that interfere destructively with certain wavelengths.

These bandgaps allow metamaterials to act as filters, blocking specific frequencies while allowing others to pass through. In photonic and phononic crystals, which are closely related to metamaterials, bandgaps enable the design of waveguides, resonators, and filters. By engineering the size, shape, and arrangement of unit cells, researchers can tune these bandgaps for precise applications in telecommunications, soundproofing, and vibration control.

1.5.5 Resonant Behavior and Localized Modes

Metamaterials often rely on resonant structures within their unit cells to achieve their unusual properties. Split-ring resonators, Helmholtz cavities, and membrane oscillators are examples of structures that interact strongly with waves at specific frequencies. These resonant behaviors amplify local fields, enabling phenomena such as negative parameters and extraordinary transmission. Resonances also give rise to localized modes, where waves become confined within certain regions of the metamaterial. These localized modes can trap energy, enhance sensitivity in sensors, and improve energy harvesting efficiency. Resonance is therefore a cornerstone in the design and operation of metamaterials across multiple domains.

1.5.6 Negative Effective Parameters

Perhaps the most striking property of metamaterials is the ability to achieve negative effective parameters. In electromagnetic metamaterials, this includes negative permittivity and permeability. In acoustic metamaterials, negative mass density and bulk modulus can be engineered. In mechanical systems, negative stiffness and compressibility have been demonstrated. When both parameters are negative simultaneously, the metamaterial is referred to as double-negative, producing highly unusual behaviors such as backward wave propagation. Negative parameters are central to applications such as negative refraction, cloaking, and the creation of super-resolution imaging systems.

1.5.7 Near-Zero Index and Extreme Parameter Materials

Beyond negative values, metamaterials can also achieve near-zero parameters, such as near-zero refractive index. Materials with near-zero indices cause waves to propagate with extremely long wavelengths inside them, effectively creating uniform phase distributions. This leads to unique effects such as tunneling of waves through narrow channels, phase front shaping, and enhanced directivity in antennas. Near-zero index materials are valuable in optics and photonics, where precise control of wavefronts is necessary for advanced communication systems. Similarly, extreme parameter metamaterials with very high or very low values expand the design space for unprecedented control over wave phenomena.

1.5.8 Nonlinear Responses and Tunability

Another fundamental behavior of metamaterials lies in their ability to exhibit nonlinear responses. While most natural materials interact with waves in linear ways, metamaterials can be designed to introduce nonlinear effects such as harmonic generation, frequency mixing, or amplitude-dependent refractive indices. These nonlinearities are achieved by integrating active components or materials that respond to external stimuli. Nonlinear metamaterials are particularly useful for dynamic and tunable applications where wave properties must be adjusted in real time. This adaptability is central to the development of reconfigurable devices, smart sensors, and active cloaking systems.

1.5.9 Losses and Dispersion

While metamaterials offer extraordinary possibilities, they also exhibit inherent challenges, such as losses and dispersion. Losses occur because resonant structures often absorb part of the wave energy, converting it into heat. This reduces efficiency and limits performance in practical applications. Dispersion refers to the dependence of metamaterial properties on frequency, meaning that unusual behaviors like negative refraction may only occur within narrow frequency ranges. Researchers are addressing these issues by developing low-loss materials, hybrid structures, and active systems that compensate for absorption. Understanding and mitigating losses and dispersion is essential for realizing the full potential of metamaterials.

1.5.10 Cloaking and Invisibility Behavior

Cloaking is perhaps the most popularized behavior of metamaterials. By carefully designing anisotropic and gradient-index structures, metamaterials can bend waves smoothly around an object, making it undetectable. Cloaking has been demonstrated not only in electromagnetic waves but also in acoustic and thermal systems. While perfect cloaking across all frequencies and conditions remains an elusive goal, partial cloaking and camouflage applications have already been achieved. Cloaking illustrates the ultimate capability of metamaterials to control wave behavior in ways that transcend the limits of natural physics.

1.5.11 Energy Focusing and Wave Guiding

Metamaterials also exhibit behaviors related to energy focusing and wave guiding. By concentrating wave energy into specific regions, metamaterials can enhance sensing, imaging, and energy harvesting systems. Similarly, waveguiding metamaterials allow waves to follow predetermined paths without scattering, even around sharp bends. These capabilities are particularly valuable in telecommunications, where controlling light or sound with minimal loss is essential for efficient data transmission. Energy focusing also plays a role in medical technologies, where acoustic metamaterials can concentrate ultrasound waves for targeted therapies.

CHAPTER 2

Theoretical Foundations

2.1 MAXWELL'S EQUATIONS AND ELECTROMAGNETIC RESPONSE

Maxwell's equations provide the bedrock upon which our understanding of electromagnetic phenomena is built, and they remain as relevant today as when they were first formulated. They describe how electric and magnetic fields originate, how they propagate, and how they respond to charges, currents, and materials. When coupled with constitutive relations, they explain the electromagnetic response of matter across the entire spectrum, from radio waves to visible light and beyond.

Their predictions encompass classical behaviors such as reflection and refraction, as well as extraordinary effects like negative refraction and cloaking in metamaterials. They also extend naturally to nonlinear and dispersive systems, highlighting their versatility and universality. The profound connection between Maxwell's equations and electromagnetic response not only revolutionized physics in the nineteenth century but continues to inspire technological innovations in optics, telecommunications, energy, and advanced materials in the twenty-first century.

2.1.1 Maxwell's Equations in Integral and Differential Form

Maxwell's equations can be expressed in both integral and differential forms, each providing unique insights into electromagnetic behavior. In their integral form, they describe how fields behave over finite regions of space, while the differential form highlights the local relationships between fields at a point. Gauss's law for electricity states that the electric flux through a closed surface is proportional to the charge enclosed, demonstrating how charges act as sources of electric fields.

Gauss's law for magnetism establishes that there are no magnetic monopoles, as the magnetic flux through a closed surface is always zero. Faraday's law of induction reveals that a changing magnetic field induces an electric field, forming the basis of electricity generation. Finally, Ampère-Maxwell's law relates magnetic fields to electric currents and includes Maxwell's correction, which introduces the displacement current, ensuring the continuity of current and enabling the prediction of electromagnetic waves. Together, these equations describe the complete behavior of classical electromagnetic fields.

2.1.2 Electromagnetic Waves and Their Propagation

One of the most profound implications of Maxwell's equations is the prediction of electromagnetic waves. By combining Faraday's law and the Ampère-Maxwell law, Maxwell demonstrated that time-varying electric and magnetic fields propagate together as coupled waves that travel through space at the speed of light. This theoretical discovery revealed that light itself is an electromagnetic wave, unifying optics with electromagnetism. The propagation of these waves depends on the electromagnetic properties of the medium, such as permittivity and permeability. In free space, the waves travel at the universal constant speed of light, while in materials, the response is modified by interactions with charges and dipoles within the medium. This modification of wave propagation underlies phenomena such as refraction, reflection, absorption, and dispersion.

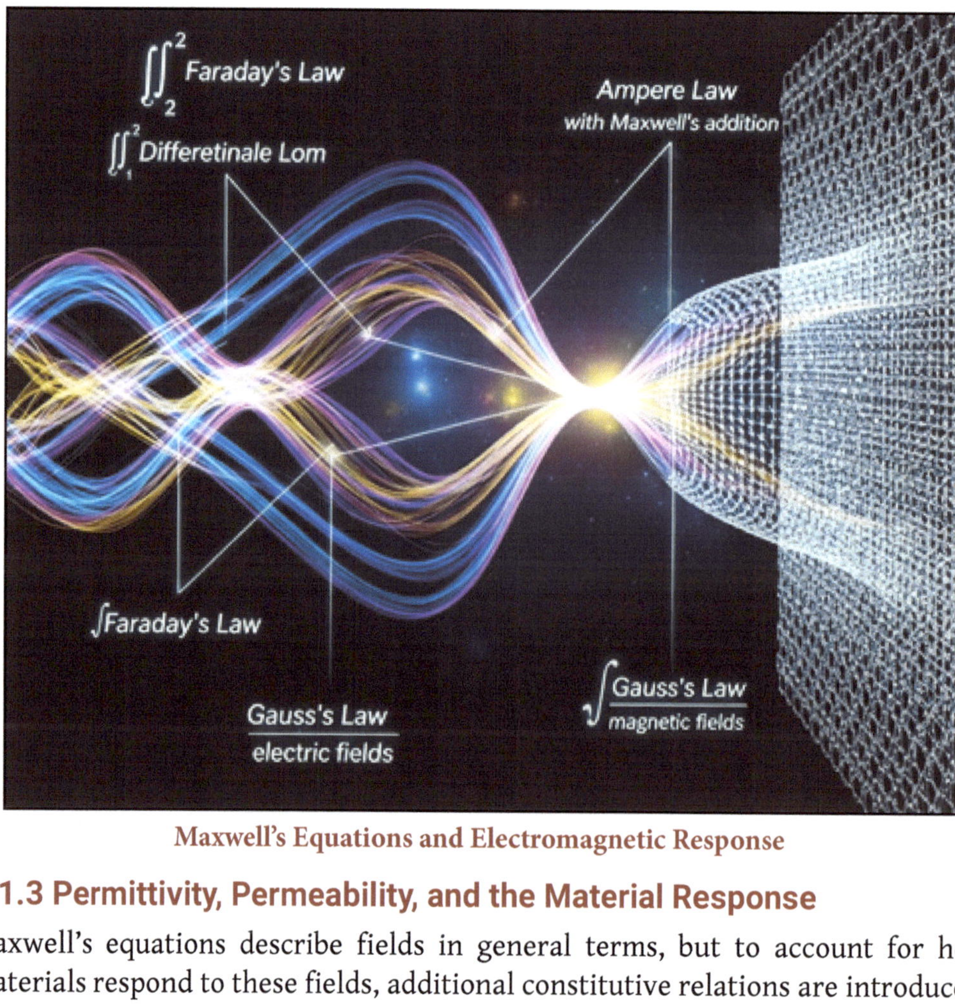

Maxwell's Equations and Electromagnetic Response

2.1.3 Permittivity, Permeability, and the Material Response

Maxwell's equations describe fields in general terms, but to account for how materials respond to these fields, additional constitutive relations are introduced.

These relations link the electric displacement field to the electric field through permittivity and the magnetic field to the magnetic flux density through permeability. Permittivity represents how strongly a material polarizes in response to an electric field, while permeability describes how it magnetizes in response to a magnetic field. Conductivity, another important property, governs how free charges move in response to electric fields, leading to currents and energy dissipation. The interplay between these parameters defines the overall electromagnetic response of a material. In metamaterials, these constitutive relations are artificially engineered by structuring materials at subwavelength scales, creating responses not observed in naturally occurring substances.

2.1.4 Boundary Conditions and Interfaces

Maxwell's equations also determine how electromagnetic fields behave at the boundaries between different materials. At an interface, the continuity conditions derived from the equations dictate how electric and magnetic field components are transmitted or reflected. The tangential components of the electric and magnetic fields must be continuous across a boundary, while the normal components of electric displacement and magnetic flux density depend on the presence of surface charge or surface current. These boundary conditions are crucial in understanding reflection and refraction phenomena, waveguides, antennas, and resonators. They also explain why metamaterials with specially designed surfaces can control electromagnetic waves in extraordinary ways, such as bending them in unconventional directions or minimizing reflection.

2.1.5 Dispersion and Frequency Dependence

Another aspect of electromagnetic response derived from Maxwell's framework is the frequency dependence of material parameters, known as dispersion. Materials respond differently to fields depending on the frequency of the incident wave. For instance, in dielectrics, polarization mechanisms such as electronic, ionic, and dipolar responses dominate at different frequency ranges. Metals exhibit strong conductivity at low frequencies but become partially transparent at optical frequencies due to the plasma frequency effect. These dispersive behaviors are described mathematically by frequency-dependent permittivity and permeability. Maxwell's equations accommodate this frequency dependence, and when combined with models like the Drude or Lorentz models, they provide powerful tools for predicting electromagnetic response across the spectrum.

2.1.6 Resonance and Localized Field Effects

Maxwell's equations also allow the description of resonance phenomena in materials and structures. When the frequency of an external electromagnetic field matches the natural oscillation frequency of charges or dipoles in a medium, resonance occurs, leading to enhanced absorption and field localization. Split-ring resonators in metamaterials are a prime example of engineered structures that exploit resonance to achieve negative permeability. Resonant interactions lead to highly localized fields, which enhance nonlinear effects, increase sensitivity in sensors, and enable strong coupling between waves and matter. These effects demonstrate how Maxwell's equations extend beyond simple wave propagation to explain complex interactions within structured materials.

2.1.7 Negative Refraction and Unconventional Behaviors

By carefully manipulating the parameters that appear in Maxwell's equations, researchers have demonstrated phenomena that were once considered impossible, such as negative refraction. When both permittivity and permeability are engineered to be negative, the resulting refractive index becomes negative, leading to wave propagation where energy and phase move in opposite directions. Maxwell's framework fully supports these results when applied with the appropriate constitutive relations, proving its universal validity. These unconventional behaviors highlight the flexibility of Maxwell's equations in describing both natural and artificial electromagnetic responses.

2.1.8 Nonlinear Electromagnetic Response

While Maxwell's equations are linear in their fundamental form, materials often exhibit nonlinear responses when subjected to strong electromagnetic fields. In nonlinear media, the polarization of the material is not directly proportional to the applied electric field but includes higher-order terms. This gives rise to effects such as harmonic generation, frequency mixing, and self-focusing of light. Nonlinear extensions of Maxwell's equations provide the foundation for nonlinear optics, which has applications in laser technology, telecommunications, and signal processing. In engineered metamaterials, nonlinear elements can be integrated into unit cells, creating systems whose response depends dynamically on the strength of the input fields.

2.1.9 Energy, Power Flow, and Conservation Laws

Maxwell's equations are also linked to conservation principles through the Poynting theorem, which describes the flow of electromagnetic energy. The Poynting vector represents the directional energy flux of the field, while the theorem ensures that the rate of energy transfer in a system is balanced by changes in stored energy and dissipated power. This connection between fields and energy emphasizes the physical reality of electromagnetic waves as carriers of energy and momentum. In practical applications, understanding energy flow is critical in designing efficient antennas, optical fibers, and energy harvesting devices.

2.2 EFFECTIVE MEDIUM THEORY (EMT)

Effective Medium Theory is a powerful and versatile tool for understanding and designing heterogeneous materials. By replacing complex, multi-component systems with equivalent homogeneous media, EMT simplifies the analysis of electromagnetic, mechanical, thermal, and acoustic responses. From classical Maxwell Garnett and Bruggeman formulations to modern extensions accounting for anisotropy, nonlinearity, and frequency dependence, EMT continues to provide essential guidance in the development of advanced composites and metamaterials. While it has limitations related to interaction effects and localized phenomena, EMT remains indispensable in both theoretical analysis and practical engineering. Its ability to connect microscopic structure with macroscopic behavior ensures that Effective Medium Theory will continue to play a central role in the study and design of next-generation materials.

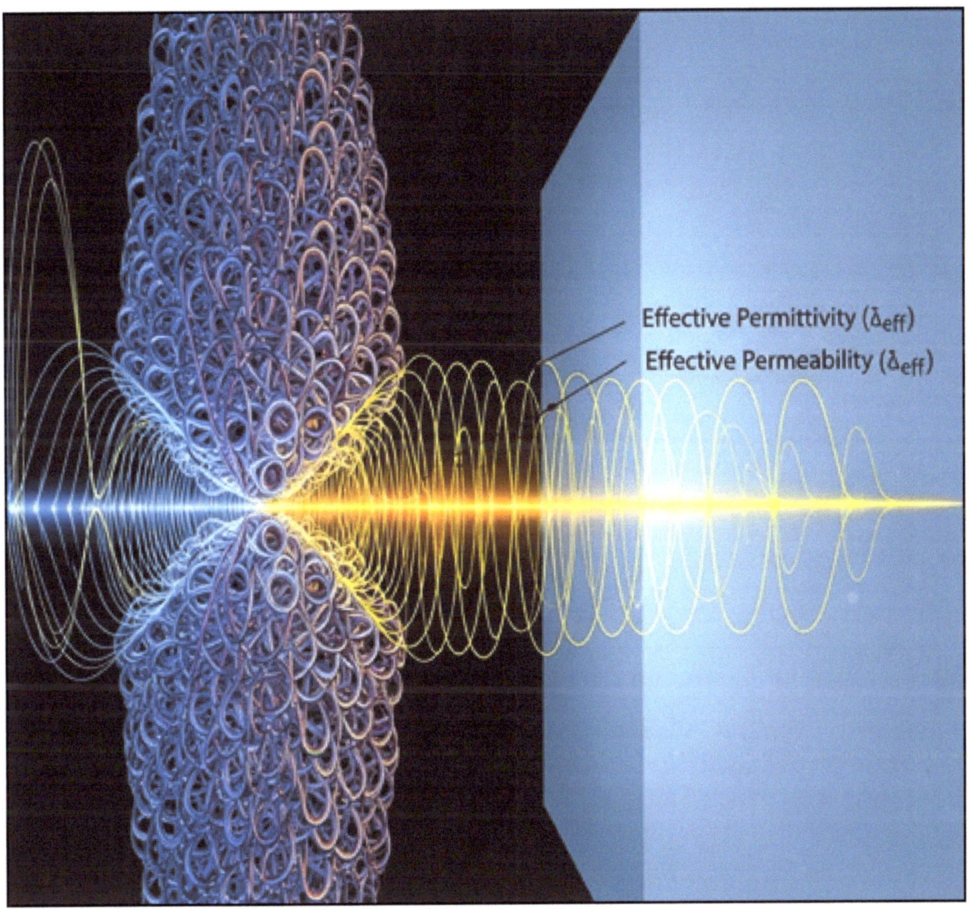

Effective Medium Theory (EMT)

2.2.1 Historical Development of EMT

The origins of Effective Medium Theory date back to the early 20th century, when researchers began investigating the macroscopic properties of mixtures and suspensions. Classical models, such as the Maxwell Garnett theory and Bruggeman approximation, emerged as early attempts to predict the effective dielectric or magnetic response of composite materials. Maxwell Garnett's formulation, developed in 1904, considered a host medium containing dilute inclusions and derived expressions for the effective permittivity by assuming that the inclusions do not interact with one another. Bruggeman's model, introduced later, extended this approach to higher volume fractions and more symmetric treatment of constituents. Over the decades, EMT has evolved to incorporate more complex microstructures, anisotropy, percolation effects, and nonlinear interactions, enabling its application to modern materials, including nanocomposites and metamaterials.

2.2.2 Basic Principles of EMT

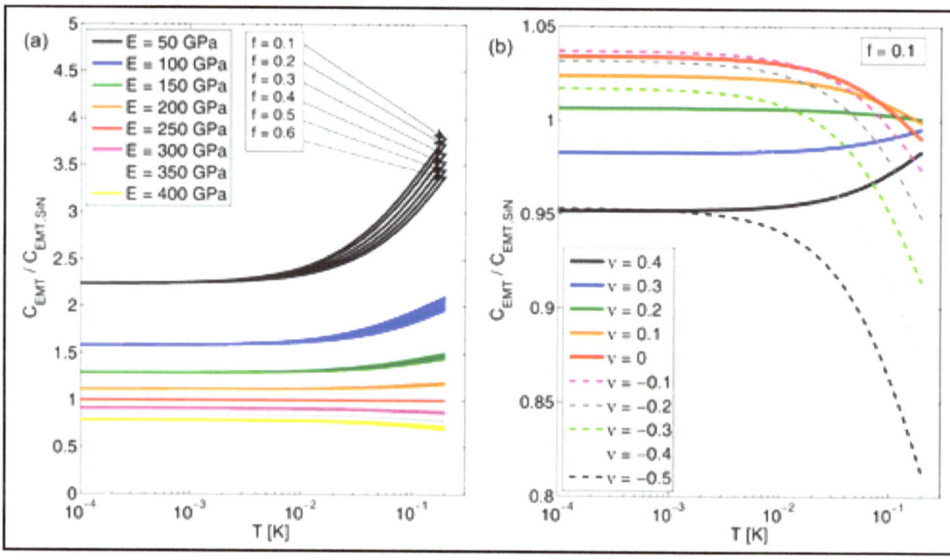

Effective Medium Theory for Low-Temperature Heat Capacity of a Metasolid Plate

The central principle of Effective Medium Theory is the replacement of a complex, heterogeneous structure with an equivalent homogeneous medium that exhibits the same macroscopic response. For electromagnetic applications, this involves averaging the local electric and magnetic fields over a representative volume, taking into account the contributions of each constituent material. The effective permittivity or permeability of the medium is then defined such that, when subjected to an external field, it produces the same average polarization or magnetization as the real heterogeneous system. In mechanical and thermal contexts, the same principle applies, where effective elastic moduli, thermal conductivities, or densities are calculated by averaging the local responses of the constituent materials. The key advantage of EMT is that it reduces a complicated, multi-scale problem into a simpler description that is analytically tractable and physically meaningful.

2.2.3 Maxwell Garnett Approximation

The Maxwell Garnett approximation is one of the earliest and most widely used formulations of EMT, particularly in the study of electromagnetic composites. In this model, a dilute concentration of inclusions with a given permittivity is embedded in a continuous host medium with a different permittivity.

The effective permittivity of the composite is determined by considering the polarization of each inclusion in response to the local electric field, under the assumption that interactions between inclusions are negligible. This approximation works well for materials with low inclusion volume fractions and when the inclusions are much smaller than the wavelength of the incident electromagnetic wave. It has been widely applied in designing dielectric composites, colloidal suspensions, and certain types of metamaterials.

2.2.4 Bruggeman Approximation

The Bruggeman approximation generalizes the Maxwell Garnett theory to composites where the constituent materials are present in comparable volume fractions. Unlike Maxwell Garnett, which treats the host and inclusion asymmetrically, Bruggeman's model treats all constituents on equal footing, considering the effective medium as a self-consistent mixture. This approach is particularly useful when dealing with percolation thresholds, where one constituent forms a continuous network within the composite. The Bruggeman model is widely applied in heterogeneous dielectrics, magnetic composites, and conductive mixtures, providing more accurate predictions in systems with higher complexity and volume fractions.

2.2.5 Anisotropic and Nonlinear Extensions

While classical EMT assumes isotropic and linear behavior, real materials often exhibit anisotropy, nonlinear responses, or frequency-dependent behavior. Modern extensions of EMT incorporate tensorial effective parameters to account for anisotropy, allowing the calculation of directional-dependent responses in layered structures or aligned composites.

Nonlinear EMT further extends the theory to materials whose response depends on the strength of the applied field, such as nonlinear dielectric or magnetoelastic composites. Frequency-dependent EMT incorporates dispersion effects, allowing the prediction of effective permittivity and permeability across a range of wavelengths. These extensions are essential for the design of advanced metamaterials, nanocomposites, and multifunctional materials.

2.2.6 Application to Metamaterials

Effective Medium Theory is particularly significant in the field of metamaterials, where the macroscopic properties arise from the engineered microstructure rather than the intrinsic properties of the base materials. In electromagnetic metamaterials, unit cells such as split-ring resonators, wire arrays, or dielectric cubes are subwavelength in size, and EMT allows their collective behavior to be described using effective permittivity and permeability. By tuning the geometry and arrangement of these unit cells, designers can achieve negative permittivity, negative permeability, near-zero index, or other exotic properties predicted by EMT. Similarly, in acoustic or mechanical metamaterials, EMT is used to calculate effective mass density, bulk modulus, or elastic modulus, guiding the design of materials with negative density, negative stiffness, or programmable elasticity.

2.2.7 Limitations and Considerations

Despite its utility, EMT has limitations that must be carefully considered. Classical EMT assumes that the inclusions are small relative to the wavelength of interest and that interactions between them are weak. When these assumptions are violated, such as in systems with large inclusions, strong coupling, or near resonances, EMT may provide inaccurate predictions. Additionally, EMT is inherently a macroscopic theory, averaging over the microscopic details, which means it cannot capture localized field enhancements, edge effects, or wave interference patterns. For highly ordered or strongly resonant structures, full-wave numerical simulations or more sophisticated homogenization techniques are often required to complement EMT.

2.2.8 Experimental Validation

The predictions of Effective Medium Theory are widely validated through experiments. In electromagnetic composites, measurements of reflection, transmission, and absorption are compared with EMT predictions to determine effective permittivity and permeability. In mechanical or thermal systems, stress-strain tests and thermal conductivity measurements verify the calculated effective elastic or thermal properties. Experimental validation has confirmed the accuracy of EMT for a wide range of materials, especially when inclusion sizes are subwavelength and volume fractions remain moderate. These validations reinforce the practical significance of EMT in both research and engineering applications.

2.2.9 Broader Implications

Beyond practical calculations, Effective Medium Theory offers fundamental insights into the nature of heterogeneous systems. It provides a unifying framework for understanding how microscopic structures determine macroscopic responses, bridging the gap between atomic-scale interactions and continuum-level behavior. EMT is not limited to classical physics but also finds applications in quantum composites, photonic crystals, and nanoscale devices. Its principles inform the design of materials with tailored optical, thermal, acoustic, and mechanical properties, highlighting the interplay between structure and function that is central to modern materials science.

2.3 NEGATIVE REFRACTIVE INDEX AND LEFT-HANDED MATERIALS

Negative refractive index and left-handed materials represent a paradigm shift in the understanding and manipulation of electromagnetic waves. By reversing conventional behaviors through engineered permittivity and permeability, LHMs enable extraordinary phenomena such as backward wave propagation, superlensing, and electromagnetic cloaking.

The theoretical foundations laid by Veselago and the experimental realization through metamaterials have transformed these concepts from speculation to practical technologies. While challenges related to fabrication, losses, and dispersion remain, continued research in this area promises to expand the capabilities of electromagnetic materials, influencing optics, telecommunications, sensing, and advanced material design. The study of LHMs exemplifies how artificial structuring at subwavelength scales can redefine the fundamental limits of wave-matter interaction, opening new frontiers in science and engineering.

The future of negative refractive index and left-handed materials lies in extending their properties to higher frequencies, including visible light and beyond. Advances in nanofabrication, 3D printing, and material synthesis are enabling increasingly complex meta-structures capable of controlling electromagnetic waves with high precision. Active and tunable LHMs, which can adapt their refractive index dynamically in response to external stimuli, are under investigation for reconfigurable lenses, adaptive antennas, and dynamic cloaking devices. Moreover, interdisciplinary research is exploring analogous concepts in acoustics, thermal conduction, and mechanical systems, where "left-handed" behaviors can manipulate sound, heat, or mechanical waves in novel ways.

Fundamentals of Refractive Index

The refractive index of a material describes how an electromagnetic wave propagates through it, defined as the ratio of the speed of light in vacuum to the phase velocity in the material. In ordinary, or right-handed, materials, the refractive index is positive, and light bends according to Snell's law. Positive refractive index materials also have permittivity and permeability that are both positive, resulting in phase and group velocities that point in the same direction. The concept of negative refractive index challenges this conventional understanding by introducing media in which both permittivity and permeability are engineered to be negative over specific frequency ranges. In such cases, the phase velocity of the wave is antiparallel to the direction of energy transport, producing unusual behaviors like backward wave propagation.

Left-Handed Materials and their Definition

Left-handed materials, also known as metamaterials with negative refractive index, derive their name from the orientation of the electric field, magnetic field, and wave vector, which form a left-handed coordinate system. In natural materials, these vectors form a right-handed system, consistent with conventional electromagnetic propagation. In LHMs, this orientation is reversed, resulting in a negative refractive index. Left-handed materials are not found naturally and must be artificially engineered. Their properties arise from periodic subwavelength structures, often referred to as meta-atoms, which collectively produce negative permittivity and permeability at certain frequencies. These engineered structures allow for precise control of electromagnetic waves, surpassing the limitations of natural substances.

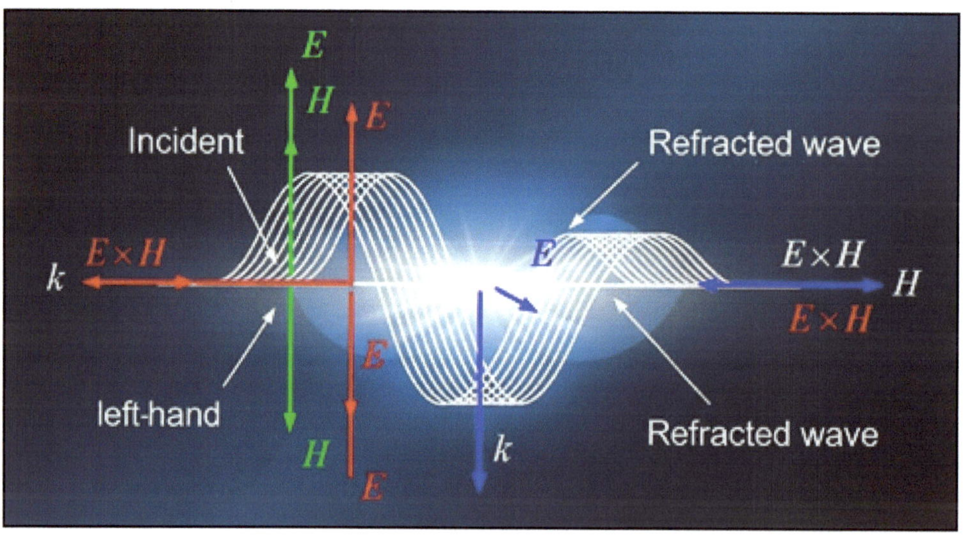

Materials with a Negative Refractive Index

Historical Development

The theoretical foundation for negative refractive index materials was laid by Viktor Veselago in 1968, who predicted the unusual electromagnetic properties of hypothetical media with negative permittivity and permeability. Veselago's work was purely theoretical at the time, as no natural material exhibited these characteristics. The experimental realization of LHMs became possible decades later with the advent of metamaterials, engineered composites designed at subwavelength scales. In 2000, experimental demonstrations of negative refractive index at microwave frequencies confirmed Veselago's predictions, employing structures such as split-ring resonators and metallic wire arrays to achieve negative permittivity and permeability. This milestone initiated a surge of research into applications and extensions across optical, terahertz, and acoustic regimes.

Electromagnetic Properties of LHMs

Left-handed materials exhibit extraordinary electromagnetic properties that differentiate them from conventional materials. When an electromagnetic wave enters an LHM, Snell's law applies, but with a negative sign, resulting in light bending in the opposite direction at an interface. The phase velocity points backward relative to the direction of energy transport, while the Poynting vector, representing energy flow, remains forward. LHMs also support reversed Doppler and Cherenkov effects, where frequencies and radiation patterns behave opposite to conventional expectations. Additionally, these materials exhibit reversed refraction at interfaces, enabling phenomena like perfect lensing, where subwavelength features are transmitted without diffraction limitations.

Design Principles and Meta-Atoms

The unique properties of left-handed materials arise from their subwavelength structural design rather than their chemical composition. Common elements include split-ring resonators, metallic rods, and periodic arrays that collectively produce effective negative permittivity and permeability. Split-ring resonators generate strong magnetic responses, while wire arrays provide negative permittivity. By carefully tuning the geometry, spacing, and orientation of these structures, engineers can control the frequency range over which the material exhibits negative refractive index. The design of meta-atoms allows for scalable and frequency-specific applications, making LHMs versatile across microwave, terahertz, and optical domains.

Superlensing and Imaging Applications

One of the most striking applications of negative refractive index materials is the concept of a superlens. Conventional lenses are limited by the diffraction limit, which restricts the resolution of imaging systems to roughly half the wavelength of light. LHMs, due to their ability to amplify evanescent waves that decay rapidly in conventional media, can overcome this limit and produce images with subwavelength resolution. Superlenses have profound implications for microscopy, nanolithography, and optical sensing, allowing for the observation and manipulation of structures at the nanoscale that were previously inaccessible.

Cloaking and Wave Manipulation

Another transformative application of left-handed materials is electromagnetic cloaking. By carefully designing the spatial variation of permittivity and permeability, LHMs can guide light around an object, making it effectively invisible to detection. Cloaking relies on transformation optics, which applies coordinate transformations to Maxwell's equations, allowing precise control of wave trajectories. Beyond invisibility, LHMs enable advanced wave manipulation such as beam steering, wavefront shaping, and antenna miniaturization. These capabilities have applications in defense, telecommunications, and sensing technologies, demonstrating the practical impact of negative refractive index materials.

Challenges and Limitations

Despite their promise, LHMs face several technical challenges. Most early demonstrations were limited to microwave frequencies, as scaling the meta-atom structures to optical wavelengths is difficult due to fabrication constraints and material losses. Additionally, resonant elements in LHMs introduce absorption, reducing efficiency and limiting practical applications. Dispersion is another concern, as negative refractive index is often achieved over narrow frequency bands, requiring precise design to match operational needs. Researchers continue to develop low-loss dielectric metamaterials, tunable structures, and hybrid composites to overcome these limitations and expand the operational bandwidth of LHMs.

2.4 PHOTONIC BAND GAPS AND RESONANCE EFFECTS

Photonic band gaps and resonance effects constitute a cornerstone of modern photonics and metamaterials research. By exploiting periodic structures and resonant interactions, it is possible to control the propagation, confinement, and localization of light with unprecedented precision. PBGs prevent propagation within specific frequency ranges, while resonances enable strong field enhancement and tunable optical responses. Together, they form the basis for innovative devices in optical communication, sensing, lasers, slow-light systems, and metamaterial applications. The interplay between structure, material properties, and wave behavior demonstrates the profound potential of photonic band gaps and resonance effects to redefine the limits of light manipulation, shaping the future of photonics, optical engineering, and advanced materials science.

2.4.1 Fundamentals of Photonic Band Gaps

A photonic band gap is a range of electromagnetic frequencies in which the propagation of light is forbidden within a periodic dielectric structure. These gaps result from the constructive and destructive interference of light waves scattered by the periodic arrangement of materials with differing refractive indices. When a wave's wavelength matches the periodicity of the structure, destructive interference prevents its propagation, creating a photonic band gap. The concept of PBGs was first theorized in the late 1980s, inspired by the electronic band theory of solids. By exploiting the analogy between electrons in crystals and photons in dielectric lattices, researchers realized that periodic dielectric structures could be engineered to control the flow of light with similar precision to how semiconductors control electron movement.

Controlling Light with Band Gaps and Resonance

2.4.2 Periodic Structures and Bragg Scattering

The formation of photonic band gaps is closely related to Bragg scattering. In a periodic medium, light reflects from successive layers, and when the path difference corresponds to an integer multiple of the wavelength, constructive interference occurs for some frequencies, while destructive interference occurs for others. The frequencies experiencing destructive interference are prohibited from propagating through the medium, forming the band gap. The size, shape, and dielectric contrast of the periodic elements determine the width and position of the band gap. Photonic crystals, which are engineered structures exhibiting periodicity in one, two, or three dimensions, are commonly used to create PBGs. The ability to control light propagation using structural periodicity allows for highly selective manipulation of electromagnetic waves.

2.4.3 Resonance Effects in Photonic Structures

Resonance effects occur when electromagnetic waves interact with structural features at frequencies that match the natural resonances of those elements. In photonic crystals and metamaterials, resonant features such as cavities, rods, or split-ring resonators trap electromagnetic energy, leading to localized field enhancement.

These resonances can significantly alter the effective permittivity and permeability of the structure, producing phenomena such as negative refractive index, slow-light propagation, and enhanced nonlinear interactions. The combination of PBGs and resonance effects enables the creation of highly tunable devices, as the resonance frequency can be adjusted by modifying the geometry or dielectric properties of the constituent elements.

2.4.4 Types of Photonic Band Gaps

Photonic band gaps can be classified based on the dimensionality of the periodic structure. One-dimensional PBGs are realized in multilayer dielectric films, where alternating layers of high and low refractive index create frequency ranges that prohibit propagation. Two-dimensional PBGs are achieved in planar photonic crystal slabs, with periodicity in two directions, commonly used in integrated photonic circuits and waveguides.

Three-dimensional PBGs require full periodicity in all spatial directions, providing complete control over light propagation in any direction and enabling applications such as optical cavities, high-Q resonators, and 3D photonic crystal fibers. The dimensionality directly influences the completeness of the band gap and the range of frequencies affected, making design considerations critical for practical applications.

2.4.5 Defects and Localized Modes

Introducing defects into photonic crystals creates localized modes within the photonic band gap. These defects act as microcavities or waveguides, trapping light at specific frequencies while still preventing the propagation of other wavelengths. The defect-induced resonances enhance light-matter interactions, making them valuable for applications in lasers, sensors, and nonlinear optical devices. By precisely controlling the location and properties of defects, researchers can engineer highly selective frequency responses, guiding light along desired paths and achieving strong field confinement. Localized modes arising from defects are central to the development of high-performance photonic devices.

2.4.6 Slow-Light and Dispersion Control

Resonance effects in combination with photonic band gaps enable the manipulation of group velocity, producing slow-light phenomena where the effective speed of light is significantly reduced.

Slow-light enhances the interaction time between light and matter, improving the efficiency of nonlinear processes, optical switching, and sensing. The dispersive properties of photonic crystals can be engineered to create flat bands near the edges of the photonic band gap, where the group velocity approaches zero. This precise control over dispersion is crucial for applications in optical buffering, delay lines, and enhanced light-matter coupling.

2.4.7 Nonlinear and Active Photonic Devices

Photonic band gaps and resonances facilitate the integration of nonlinear and active components into photonic devices. Nonlinear effects, such as second-harmonic generation, Kerr-induced phase shifts, and four-wave mixing, are enhanced by localized resonances, enabling low-power all-optical switching and signal processing. Active materials, including gain media and tunable dielectrics, can be incorporated into photonic structures to achieve dynamic control of the band gap or resonance frequency. These capabilities allow the creation of reconfigurable photonic devices, adaptive filters, and high-efficiency lasers, highlighting the versatility of PBGs and resonances in modern optical engineering.

2.4.8 Applications in Optical Communication

In optical communication systems, photonic band gaps and resonances provide precise control over light propagation, filtering, and confinement. Waveguides based on photonic crystals can guide light around sharp bends without significant loss, while band gap engineering ensures that unwanted frequencies are blocked. Resonant cavities serve as wavelength-selective elements for multiplexing and demultiplexing signals, improving the efficiency and capacity of communication networks. The combination of low-loss waveguiding, high-Q resonators, and tunable band gaps enables compact, high-performance photonic circuits that are integral to modern telecommunication infrastructure.

2.4.9 Metamaterials and Enhanced Light Manipulation

Photonic band gaps and resonance effects are closely linked to the field of metamaterials. By embedding resonant meta-atoms within periodic structures, designers can create negative permittivity and permeability, leading to negative refractive index and cloaking phenomena. Resonances enhance the electromagnetic response at specific frequencies, while the periodic structure introduces selective band gaps, enabling precise control of wave propagation. This synergy allows for advanced light manipulation, including superlensing, beam shaping, and highly directional emission, expanding the range of optical phenomena accessible to engineers and researchers.

2.5 SURFACE PLASMONS AND SUBWAVELENGTH EFFECTS

Surface plasmons and subwavelength effects have transformed the understanding and manipulation of light at the nanoscale. By coupling electromagnetic waves with electron oscillations at metal-dielectric interfaces, plasmons enable confinement of light beyond the diffraction limit, field enhancement, and tunable resonance phenomena. These effects underpin applications in sensing, spectroscopy, imaging, energy harvesting, and integrated plasmonic circuits. While challenges remain in terms of losses, fabrication, and integration, ongoing research into novel materials, hybrid structures, and active control mechanisms is driving the field forward. The study of surface plasmons and subwavelength optics continues to expand the limits of light-matter interaction, shaping the future of nanophotonics and advanced optical technologies.

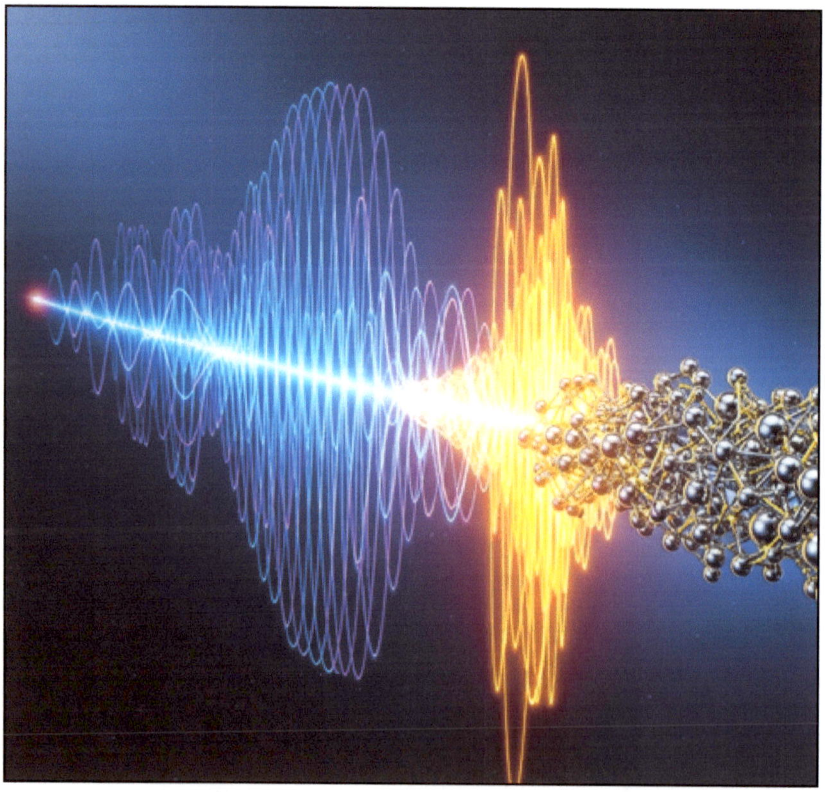

Principles of Surface Plasmons and Subwavelength Confinement

The future of surface plasmons and subwavelength optics is focused on extending their capabilities in integrated photonics, sensing, energy harvesting, and quantum technologies. Novel materials such as graphene, transition metal dichalcogenides, and topological insulators offer tunable plasmonic properties, low losses, and compatibility with flexible substrates. Active plasmonic devices, which can be modulated electrically, optically, or thermally, enable dynamic control of light at the nanoscale. Combining plasmonics with metamaterials, photonic crystals, and nonlinear materials opens new avenues for light manipulation, super-resolution imaging, and ultrafast optical processing. These developments promise to expand the technological impact of plasmonics across multiple scientific and industrial domains.

2.5.1 Fundamentals of Surface Plasmons

Surface plasmons arise due to the collective oscillation of free electrons at the boundary between a metal and a dielectric when excited by an electromagnetic field. The oscillations produce a coupled electromagnetic wave that propagates along the interface, with the electric field decaying exponentially perpendicular to it. This confinement of electromagnetic energy near the surface leads to enhanced local fields, which are highly sensitive to changes in the surrounding environment. The resonance condition for surface plasmons depends on the dielectric function of the metal, the permittivity of the adjacent dielectric, and the wave vector of the incident light. Metals such as gold, silver, and aluminum are commonly used due to their high density of free electrons and suitable optical properties, making them ideal platforms for plasmonic devices.

2.5.2 Surface Plasmon Polaritons

A key concept in surface plasmon studies is the surface plasmon polariton (SPP), a hybrid mode that combines the collective electron oscillation with an electromagnetic wave propagating along the metal-dielectric interface. SPPs are characterized by strong field confinement and dispersion that differs from conventional light waves. They exhibit wavelengths shorter than free-space light at the same frequency, enabling subwavelength control of optical fields. The propagation of SPPs is influenced by the metal's permittivity and the dielectric environment, and their attenuation is governed by absorption losses in the metal. SPPs serve as the foundation for plasmonic waveguides, sensors, and circuits that operate at the nanoscale, where conventional photonics would be limited by diffraction.

2.5.3 Localized Surface Plasmons

In addition to propagating surface plasmons, localized surface plasmons (LSPs) occur in nanostructures such as metallic nanoparticles, nanorods, or nanoshells.

LSPs result from the confinement of electron oscillations within subwavelength volumes, producing intense near-field enhancements and resonance peaks in the optical spectrum. The resonance frequency depends on the particle size, shape, composition, and the dielectric environment. Localized surface plasmons are particularly useful in sensing applications, as small changes in the surrounding medium induce measurable shifts in the resonance wavelength. They also play a crucial role in surface-enhanced spectroscopies, such as surface-enhanced Raman scattering (SERS), where the enhanced electromagnetic fields amplify molecular signals.

2.5.4 Subwavelength Confinement of Light

One of the most significant consequences of surface plasmon excitation is the ability to confine light to dimensions much smaller than its free-space wavelength. This subwavelength confinement overcomes the traditional diffraction limit, which restricts the resolution and miniaturization of optical devices.

Plasmonic nanostructures, including nanoantennas, nanoparticle arrays, and waveguides, exploit these effects to concentrate light into nanoscale volumes, enhancing field intensities and enabling strong light-matter interactions. Subwavelength confinement is critical for applications in high-resolution imaging, on-chip photonics, optical trapping, and enhanced nonlinear optical processes.

2.5.5 Resonance Effects and Field Enhancement

Surface plasmons exhibit strong resonance effects, where the incident light couples efficiently to electron oscillations at specific frequencies. At resonance, the local electric field near the metal surface can be amplified by several orders of magnitude, providing a powerful tool for enhancing optical signals. These resonance effects depend sensitively on the nanostructure geometry and the surrounding dielectric environment, allowing precise tuning of the plasmonic response. Resonant plasmonic structures are essential for sensors, where even minute changes in refractive index produce significant shifts in resonance wavelength, enabling ultra-sensitive detection of biomolecules, chemicals, or environmental changes.

2.5.6 Plasmonic Waveguides and Circuits

Surface plasmons provide a platform for guiding light in structures much smaller than the wavelength of light, giving rise to plasmonic waveguides and integrated circuits. By engineering the shape and arrangement of metallic nanostructures, it is possible to direct SPPs along defined paths, bend them around sharp corners, and couple them into other photonic or electronic components.

Plasmonic circuits offer the potential for ultrafast, nanoscale optical communication, bridging the gap between conventional electronics and photonics. The strong confinement and field enhancement inherent to plasmons allow for high-density integration and low-power operation, making them promising candidates for future optical computing and on-chip signal processing.

2.5.7 Sensing and Spectroscopy Applications

The sensitivity of surface plasmons to changes in the dielectric environment underlies numerous sensing and spectroscopy applications. Surface plasmon resonance (SPR) sensors detect minute variations in refractive index near the metal surface, making them invaluable in chemical and biological detection. Localized surface plasmon resonance (LSPR) sensors exploit nanoparticle resonances to provide enhanced sensitivity in compact configurations. In spectroscopy, plasmon-enhanced techniques such as SERS leverage the field enhancement near plasmonic nanostructures to detect single molecules, monitor chemical reactions, and study nanoscale interactions. The combination of field enhancement, subwavelength confinement, and tunable resonance makes plasmonics a versatile platform for precision sensing.

2.5.8 Energy Harvesting and Photothermal Effects

Surface plasmons are also applied in energy harvesting and photothermal applications. The strong local field and resonant absorption enable efficient conversion of light into heat, useful for photothermal therapy, localized heating, and solar energy applications. Plasmonic nanostructures can enhance light absorption in photovoltaic devices, increasing the efficiency of solar cells by trapping and concentrating light in active layers. Subwavelength effects allow for optimized light-matter interaction in thin films or nanoscale absorbers, reducing material usage while maximizing energy conversion. These applications illustrate the broad technological relevance of surface plasmons and subwavelength optics.

2.5.9 Challenges

Despite their potential, surface plasmons and subwavelength effects face challenges related to losses, fabrication, and integration. Metals used in plasmonics, such as gold and silver, exhibit intrinsic absorption losses that limit the propagation length of SPPs and reduce device efficiency. Fabricating nanoscale structures with the required precision is demanding and requires advanced lithography or self-assembly techniques. Additionally, integrating plasmonic components with conventional photonic or electronic devices requires careful design to manage impedance matching, mode coupling, and thermal effects. Ongoing research focuses on low-loss plasmonic materials, hybrid structures, and scalable fabrication methods to address these limitations.

CHAPTER 3

Electromagnetic Metamaterials

3.1 SPLIT-RING RESONATORS (SRRS)

Split-ring resonators are fundamental components in the field of metamaterials, enabling the realization of negative permeability, negative refractive index, and subwavelength light manipulation. Their LC-type resonance, subwavelength geometry, and field localization properties provide unprecedented control over electromagnetic waves. Arrays of SRRs amplify these effects, allowing the design of bulk metamaterials and advanced optical devices. While challenges related to fabrication, losses, and integration remain, ongoing research in tunable, reconfigurable, and hybrid SRRs continues to expand their potential. By bridging the gap between structural engineering and electromagnetic functionality, SRRs have become indispensable in the design of modern metamaterials, sensors, and nanoscale photonic devices.

Future research on SRRs is focused on integrating them with emerging materials and technologies to expand their capabilities. Hybrid SRRs combining metals with dielectrics, graphene, or phase-change materials offer tunable and reconfigurable functionality. Active control through electrical, optical, or thermal modulation enables adaptive metamaterials for dynamic applications. Combining SRRs with other metamaterial elements, such as wire arrays or plasmonic structures, allows simultaneous control of electric and magnetic responses. Additionally, research is exploring SRRs for quantum applications, nonlinear optics, and energy harvesting, demonstrating their continuing relevance in advanced photonics and metamaterial design.

3.1.1 Structure and Geometry

A split-ring resonator typically consists of a pair of concentric metallic rings with gaps or splits at opposite sides. These gaps interrupt the current path, allowing the SRR to behave as a resonant LC circuit, where the ring inductance and gap capacitance determine the resonance frequency. Variations in the geometry, such as the size, spacing, thickness, and number of splits, allow precise tuning of the resonant response. The subwavelength dimensions of SRRs are critical, as they enable effective medium approximations in metamaterials, allowing the artificial medium to interact with electromagnetic waves as if it were homogeneous. The planar or three-dimensional arrangement of SRRs in arrays further enhances their collective electromagnetic response, enabling the realization of bulk metamaterials with engineered properties.

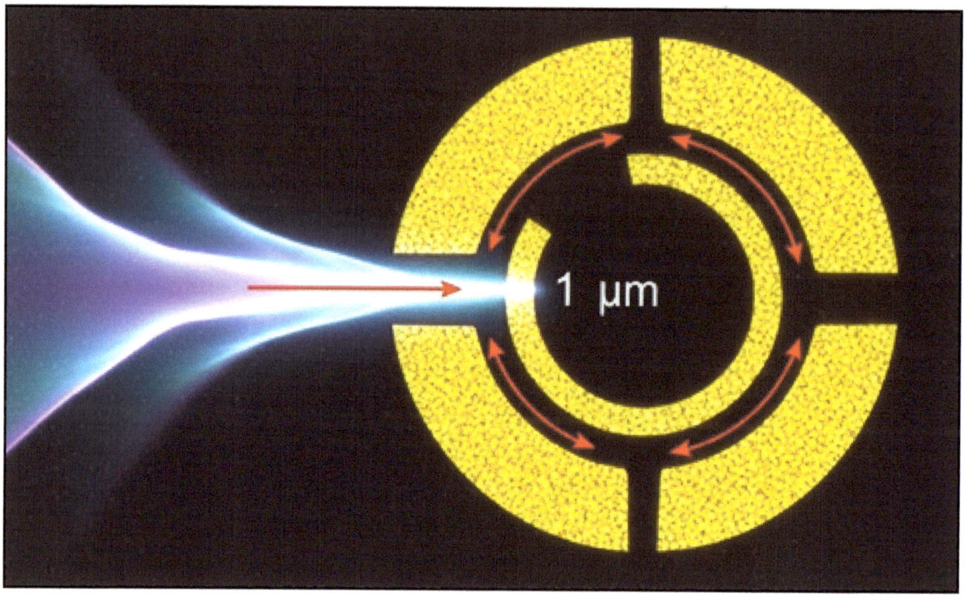

The Function and Design of Split-Ring Resonators

3.1.2 Electromagnetic Response

The unique electromagnetic response of SRRs arises from their ability to support circulating currents when exposed to a magnetic field perpendicular to the plane of the rings. These induced currents create a magnetic moment opposite to the applied field, producing an effective negative permeability over a specific frequency range. When incorporated into metamaterials with complementary electric elements, such as metallic wires providing negative permittivity, SRRs contribute to achieving negative refractive index behavior. The resonance frequency and bandwidth of SRRs depend on their geometric parameters, the dielectric environment, and the arrangement in arrays. By manipulating these factors, engineers can design metamaterials with tailored electromagnetic responses across microwave, terahertz, and optical regimes.

3.1.3 Resonance Mechanism

The operation of SRRs is analogous to an LC circuit, where the inductance is associated with the ring's loop and the capacitance arises from the gap between the ends of the rings. When an incident electromagnetic wave induces an alternating current in the SRR, the interaction between inductance and capacitance produces a resonance at a specific frequency. At resonance, the circulating current produces a magnetic field that can oppose the external field, resulting in strong magnetic activity even in non-magnetic materials. This resonance is essential for achieving negative permeability, as it allows SRRs to exhibit magnetic responses at frequencies where natural materials are non-magnetic. The resonance also leads to localized field enhancement near the splits, which can be exploited for sensing and nonlinear applications.

3.1.4 Arrays and Collective Effects

While a single SRR exhibits magnetic resonance, arranging SRRs in periodic arrays introduces collective effects that enhance and broaden the electromagnetic response. Coupling between neighboring SRRs can lead to hybridized resonances, increased field confinement, and tunable effective material parameters.

The periodicity of the array, orientation of the SRRs, and inter-element spacing determine the strength and frequency range of the collective response. These array effects are essential for realizing bulk metamaterials with homogeneous effective properties and for creating devices with tailored electromagnetic behavior, such as negative refractive index metamaterials, frequency-selective surfaces, and metamaterial absorbers.

3.1.5 Frequency Scaling and Applications

SRRs have been demonstrated across a wide range of frequencies, from microwave to terahertz and optical domains. Scaling the dimensions of the SRR elements shifts the resonance frequency, allowing engineers to design metamaterials for specific applications.

In the microwave regime, SRRs are used in antennas, filters, and compact waveguides. In the terahertz and optical ranges, they enable subwavelength light manipulation, enhanced nonlinear effects, and plasmonic applications. The ability to scale SRRs across frequency ranges makes them versatile components in electromagnetic engineering, bridging the gap between conventional materials and advanced functional metamaterials.

3.1.6 Tunable and Reconfigurable SRRs

Modern research has focused on developing tunable and reconfigurable SRRs, allowing dynamic control over their resonant behavior. By incorporating varactor diodes, phase-change materials, or micro-electromechanical systems (MEMS), the resonance frequency and effective permeability can be adjusted in real time. Tunable SRRs enable adaptive metamaterials capable of responding to environmental changes, electrical signals, or optical stimuli. These capabilities are particularly valuable for applications in adaptive filters, modulators, sensors, and beam-steering devices, where flexibility and real-time control are essential.

3.1.7 Subwavelength and Field Localization Effects

SRRs inherently operate at subwavelength scales, concentrating electromagnetic fields within and around their structure. The splits in the rings create regions of intense electric field, while the circulating currents produce strong magnetic fields. This localized field enhancement can be exploited for applications in sensing, nonlinear optics, and energy harvesting. Subwavelength confinement also enables SRRs to interact strongly with incident waves despite their small physical size, making them ideal candidates for compact metamaterial designs. The combination of subwavelength operation and resonance leads to unique electromagnetic phenomena that are inaccessible in conventional materials.

3.1.8 Sensing and Detection Applications

The strong resonance and field enhancement in SRRs make them effective tools for sensing and detection. Small changes in the surrounding dielectric environment, such as the presence of a molecule, refractive index variation, or temperature change, can shift the resonance frequency, allowing precise detection of environmental changes.

SRR-based sensors have been demonstrated for chemical, biological, and environmental monitoring, with high sensitivity and specificity. By optimizing the geometry and arrangement of SRRs, sensors can be designed to target specific frequency ranges and enhance the interaction with the analyte, making them powerful platforms for nanoscale and microscale detection.

3.1.9 Challenges

Despite their versatility, SRRs face several challenges related to fabrication, losses, and integration. At higher frequencies, especially in the optical regime, fabricating nanoscale SRRs with precise dimensions and alignment is technically demanding. Metallic losses, particularly in optical SRRs, reduce the quality factor of the resonance and limit the effectiveness of metamaterials. Coupling between SRRs in dense arrays can also introduce unwanted interference or mode splitting. Addressing these challenges requires advanced nanofabrication techniques, low-loss materials, and careful design of array geometry to optimize performance.

3.2 WIRE ARRAYS AND MAGNETIC METAMATERIALS

Wire arrays and magnetic metamaterials represent a powerful combination for engineering electromagnetic responses beyond the limits of natural materials. Wire arrays provide negative permittivity, while magnetic resonators generate negative permeability, enabling the creation of negative refractive index metamaterials with unique wave propagation characteristics. Their subwavelength operation, resonance behavior, and tunability allow the design of advanced devices for superlensing, cloaking, sensing, and telecommunications. While challenges related to losses, fabrication, and bandwidth remain, ongoing research in hybrid and reconfigurable systems continues to expand the capabilities of wire-metamaterial composites. These structures exemplify the transformative potential of metamaterials in controlling light and electromagnetic waves at the nanoscale and beyond.

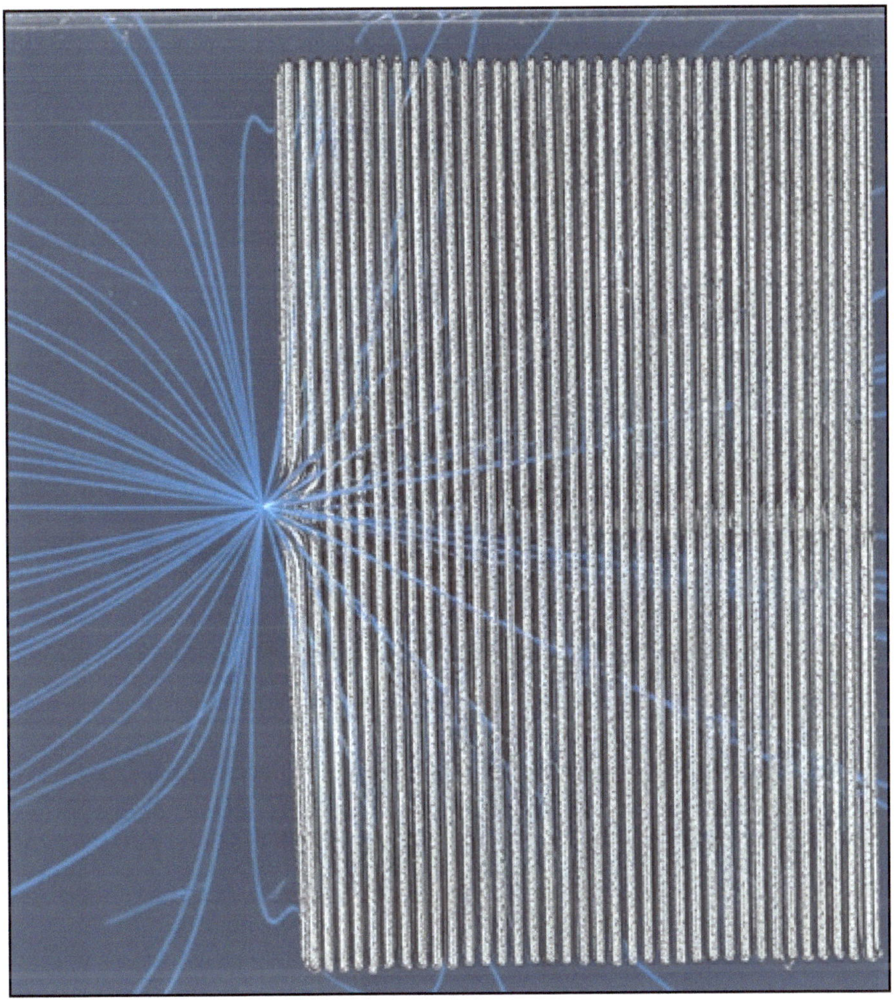

The Role of Wire Arrays in Magnetic Metamaterials

The future of wire arrays and magnetic metamaterials is focused on expanding their operational frequency ranges, reducing losses, and integrating them with emerging materials such as graphene, topological insulators, and low-loss dielectrics. Hybrid systems combining wire arrays with plasmonic, photonic crystal, or nonlinear elements promise enhanced field control, broader bandwidth, and novel functionalities. Research is also exploring their use in quantum devices, energy harvesting, and ultrafast photonics. The ability to design materials with independently controlled electric and magnetic responses continues to drive innovation in metamaterial science, pushing the boundaries of electromagnetic wave manipulation.

3.2.1 Wire Arrays: Structure and Function

Wire arrays consist of periodic metallic rods or wires embedded in a dielectric medium, arranged in one-, two-, or three-dimensional lattices. These structures interact with incident electromagnetic fields to produce an effective negative permittivity over a certain frequency range. The response of wire arrays arises from the collective motion of conduction electrons along the wires, generating plasmonic-like oscillations that mimic the behavior of free electron plasmas. The effective plasma frequency depends on the spacing, diameter, and arrangement of the wires, allowing precise tuning of the frequency range over which negative permittivity is achieved. Wire arrays are particularly useful in combination with magnetic resonators, as they provide the complementary electric response required to realize negative refractive index metamaterials.

3.2.2 Magnetic Metamaterials and Negative Permeability

Magnetic metamaterials are artificial structures designed to exhibit magnetic responses that natural materials cannot achieve at high frequencies. Conventional materials are typically non-magnetic at microwave, terahertz, and optical frequencies, but engineered elements such as split-ring resonators, spiral resonators, or fishnet structures can produce negative permeability over specific frequency bands. When arranged in periodic arrays, these resonators interact collectively to create bulk magnetic metamaterials with homogeneous effective properties. By combining negative permittivity from wire arrays and negative permeability from magnetic resonators, it becomes possible to realize left-handed materials with negative refractive index, enabling extraordinary wave phenomena such as backward wave propagation, subwavelength focusing, and cloaking.

3.2.3 Electromagnetic Response of Wire Arrays

The electromagnetic response of wire arrays is primarily determined by the conduction electrons along the wires and the periodicity of the lattice. At frequencies below the effective plasma frequency, the wires exhibit a negative permittivity, causing incident electromagnetic waves to be strongly reflected or attenuated. The collective behavior of the wires generates a plasma-like response, analogous to that seen in free electron metals, but with tunable characteristics determined by the lattice geometry. This effective permittivity can be manipulated by varying wire diameter, spacing, or the dielectric environment, allowing engineers to design metamaterials with tailored electric responses that complement magnetic resonators and achieve desired electromagnetic properties.

3.2.4 Combined Wire-Metamaterial Systems

The integration of wire arrays with magnetic metamaterials creates composite structures capable of exhibiting both negative permittivity and negative permeability. In these systems, the wire arrays provide the necessary electric response, while the magnetic resonators introduce negative permeability over overlapping frequency ranges. The combination enables the design of negative refractive index metamaterials, where the phase velocity of waves is antiparallel to the direction of energy propagation. Such materials exhibit counterintuitive electromagnetic phenomena, including reversed Snell's law, inverted Doppler effects, and enhanced wave localization. The design of these composites requires careful tuning of both electric and magnetic elements to achieve overlapping resonances and maximize the operational bandwidth.

3.2.5 Resonance and Bandwidth Considerations

The performance of wire arrays and magnetic metamaterials is closely linked to resonance effects. Magnetic resonators such as split-ring resonators produce strong magnetic responses at their resonance frequency, while wire arrays exhibit effective negative permittivity below their plasma frequency.

Achieving simultaneous negative permittivity and permeability requires overlapping these resonance bands. The bandwidth of negative refractive index materials is therefore determined by the tuning of wire geometry, resonator dimensions, and array periodicity. Coupling effects between neighboring elements can broaden or split resonance bands, providing additional flexibility for engineering the desired electromagnetic response. Advanced design strategies, including multi-resonator configurations and hybrid wire-metamaterial systems, are used to optimize bandwidth and minimize losses.

3.2.6 Subwavelength and Field Localization Effects

Wire arrays and magnetic metamaterials operate at subwavelength scales, allowing control of electromagnetic waves below the diffraction limit. The localized fields around resonators and along the wires enhance light-matter interaction, enabling strong field confinement and enhanced nonlinear effects. Subwavelength operation is essential for realizing compact metamaterial devices, including miniaturized waveguides, antennas, and sensors. Field localization in wire-metamaterial composites also enhances the sensitivity of plasmonic and electromagnetic devices, making them ideal for applications in high-resolution imaging, chemical sensing, and energy concentration at the nanoscale.

3.2.7 Applications in Negative Index Metamaterials

Wire arrays combined with magnetic metamaterials form the basis of negative refractive index metamaterials, which have transformative applications in optics and electromagnetics. These materials enable superlensing, where evanescent waves are amplified, allowing imaging below the diffraction limit. They also facilitate cloaking, where electromagnetic waves are guided around objects to render them invisible to detection. In addition, negative index metamaterials are used to design compact antennas with improved directivity, tunable filters for telecommunications, and novel waveguides for controlling light propagation in integrated photonic circuits. The precise control of electric and magnetic responses in wire-metamaterial composites underpins these revolutionary applications.

3.2.8 Tunable and Reconfigurable Systems

Recent advances have focused on making wire arrays and magnetic metamaterials tunable and reconfigurable. Incorporating varactor diodes, phase-change materials, liquid crystals, or MEMS-based structures allows dynamic control of the effective permittivity and permeability. Such tunable systems enable adaptive metamaterials capable of changing their electromagnetic response in real time in response to electrical, optical, or thermal stimuli. Reconfigurable metamaterials have applications in adaptive filtering, beam steering, and frequency-agile antennas, providing versatility that is crucial for next-generation communication and sensing technologies.

3.3 TERAHERTZ AND MICROWAVE METAMATERIALS

Terahertz and microwave metamaterials provide unprecedented control over electromagnetic waves, enabling negative permittivity, negative permeability, and subwavelength manipulation of light. Wire arrays and split-ring resonators form the backbone of these metamaterials, allowing engineered responses that extend beyond natural materials. Applications span communication systems, imaging, sensing, superlensing, and frequency-selective surfaces. While challenges remain in terms of losses, fabrication, and bandwidth, ongoing research in tunable and reconfigurable metamaterials promises versatile and adaptive devices. Terahertz and microwave metamaterials exemplify the transformative potential of engineered structures in controlling electromagnetic waves, paving the way for advanced technologies in multiple scientific and technological fields.

3.3.1 Fundamentals of Terahertz and Microwave Metamaterials

Metamaterials operate by structuring subwavelength inclusions, which act as meta-atoms to produce desired electromagnetic responses. In the microwave and terahertz regimes, the wavelength of electromagnetic waves ranges from millimeters to hundreds of micrometers, enabling the fabrication of metamaterials using lithography, printed circuit boards, and microfabrication techniques. The effective permittivity and permeability of these metamaterials are controlled by the geometry, orientation, and periodicity of the inclusions. Microwave metamaterials often use metallic structures such as split-ring resonators and wire grids, while terahertz metamaterials incorporate smaller-scale resonators, dielectric resonators, and hybrid metallic-dielectric designs. The ability to manipulate wave propagation in these frequency ranges has opened new avenues for waveguiding, filtering, and sensing applications.

Metamaterials for Terahertz and Microwave Frequencies

3.3.2 Split-Ring Resonators and Negative Permeability

Split-ring resonators (SRRs) are a cornerstone of microwave and terahertz metamaterials. SRRs consist of metallic rings with gaps that create resonant LC circuits. When exposed to a magnetic field perpendicular to the plane of the rings, circulating currents are induced, producing a magnetic response that can be negative over a specific frequency range. Negative permeability achieved through SRRs allows the design of left-handed metamaterials when combined with elements that produce negative permittivity, such as wire arrays. In the terahertz regime, SRRs must be miniaturized while maintaining precise geometric proportions to achieve resonance at shorter wavelengths. These structures are instrumental in producing tunable, frequency-selective metamaterials with engineered electromagnetic properties.

3.3.3 Wire Arrays and Negative Permittivity

Wire arrays are periodic arrangements of metallic wires that produce negative permittivity over a range of frequencies below their effective plasma frequency. The conduction electrons in the wires respond collectively to incident electromagnetic waves, creating a plasma-like response that blocks certain frequencies from propagating. In combination with SRRs, wire arrays enable the realization of metamaterials with simultaneously negative permittivity and permeability, producing negative refractive index behavior. Wire arrays in microwave metamaterials are relatively straightforward to fabricate using printed circuit techniques, while terahertz wire arrays require more precise fabrication to maintain subwavelength dimensions. The tunability of wire array parameters, including spacing, diameter, and arrangement, allows for the precise design of electromagnetic properties for specific applications.

3.3.4 Resonance and Subwavelength Effects

Resonance plays a central role in the functionality of terahertz and microwave metamaterials. Both SRRs and wire arrays exhibit resonances determined by their geometric parameters and the dielectric environment. At resonance, the electromagnetic fields are strongly confined and enhanced, producing localized energy regions that enable subwavelength manipulation of waves. These subwavelength effects are crucial for designing compact devices such as filters, waveguides, antennas, and sensors. Field localization at resonant sites allows for increased interaction between electromagnetic waves and matter, enhancing device sensitivity and enabling applications in imaging, spectroscopy, and nonlinear optics.

3.3.5 Terahertz Metamaterials: Opportunities and Challenges

Terahertz metamaterials address the "terahertz gap," a frequency range where conventional electronic devices are inefficient, and optical devices do not operate effectively. By engineering meta-atoms for resonance in the terahertz range, metamaterials enable applications in spectroscopy, imaging, and communication. Terahertz waves can penetrate materials such as clothing, plastics, and paper, making them ideal for security screening, non-destructive testing, and medical imaging. However, fabricating terahertz metamaterials poses significant challenges due to the small dimensions required for resonant structures and high material losses in metals at these frequencies. Advanced fabrication techniques, such as electron-beam lithography and nanoimprint lithography, are essential for creating functional terahertz metamaterials.

3.3.6 Microwave Metamaterials: Practical Implementations

Microwave metamaterials are more mature in terms of fabrication and applications compared to terahertz metamaterials. They are widely used in antenna miniaturization, beam shaping, electromagnetic shielding, and frequency-selective surfaces. Microwave metamaterials often employ SRRs, wire arrays, and complementary split-ring resonators to achieve desired electromagnetic properties. The longer wavelengths at microwave frequencies allow for easier fabrication and lower losses, making them suitable for real-world applications such as radar systems, satellite communications, and wireless networks. By controlling both the electric and magnetic response, engineers can design metamaterials with unique wave propagation characteristics, including negative refraction, backward wave propagation, and subwavelength focusing.

3.3.7 Tunable and Reconfigurable Metamaterials

Tunable metamaterials are an active area of research in both the microwave and terahertz regimes. Incorporating varactors, micro-electromechanical systems (MEMS), phase-change materials, or liquid crystals into metamaterial structures allows dynamic adjustment of resonant frequency, permittivity, and permeability. This tunability enables adaptive filters, modulators, frequency-selective surfaces, and beam-steering devices. Reconfigurable metamaterials provide versatility for applications where environmental conditions or operational requirements change, such as adaptive radar systems, cognitive radio networks, and responsive terahertz imaging platforms.

3.3.8 Sensing and Imaging Applications

Terahertz and microwave metamaterials enhance sensing and imaging capabilities by exploiting resonance and subwavelength field localization. Metamaterials can amplify weak signals, concentrate electromagnetic energy in small volumes, and selectively filter frequencies. In sensing applications, changes in the dielectric environment or target proximity shift the resonance frequency, enabling highly sensitive detection of chemical, biological, or structural changes. In imaging, metamaterials facilitate subwavelength resolution, enhanced contrast, and reduced scattering, making them useful in security screening, medical diagnostics, and industrial inspection. These applications highlight the practical impact of engineered metamaterials in diverse technological fields.

3.4 METAMATERIALS FOR ANTENNAS AND WAVEGUIDES

3.4.1 Metamaterials for Antennas

Metamaterial antennas are a class of antennas which use metamaterials to increase performance of miniaturized (electrically small) antenna systems. Their purpose, as with any electromagnetic antenna, is to launch energy into free space. However, this class of antenna incorporates metamaterials, which are materials engineered with novel, often microscopic, structures to produce unusual physical properties. Antenna designs incorporating metamaterials can step-up the antenna's radiated power.

Conventional antennas that are very small compared to the wavelength reflect most of the signal back to the source. A metamaterial antenna behaves as if it were much larger than its actual size, because its novel structure stores and re-radiates energy. Established lithography techniques can be used to print metamaterial elements on a printed circuit board. These novel antennas aid applications such as portable interaction with satellites, wide angle beam steering, emergency communications devices, micro-sensors and portable ground-penetrating radars to search for geophysical features. Some applications for metamaterial antennas are wireless communication, space communications, GPS, satellites, space vehicle navigation and airplanes.

This Z Antenna Tested at the National Institute of Standards and Technology is Smaller than a Standard Antenna with Comparable Properties. Its High Efficiency is Derived from the "Z Element" inside the Square that Acts as a Metamaterial, Greatly Boosting the Radiated Signal. The Square is 30 Millimeters on a Side

Antenna Designs

Antenna designs incorporating metamaterials can improve the radiated power of an antenna. The newest metamaterial antennas radiate as much as 95 percent of an input radio signal. Standard antennas need to be at least half the size of the signal wavelength to operate efficiently. At 300 MHz, for instance, an antenna would need to be half a meter long. In contrast, experimental metamaterial antennas are as small as one-fiftieth of a wavelength, and could have further decreases in size.

Metamaterials are a basis for further miniaturization of microwave antennas, with efficient power and acceptable bandwidth. Antennas employing metamaterials offer the possibility of overcoming restrictive efficiency-bandwidth limitations for conventionally constructed, miniature antennas. Metamaterials permit smaller antenna elements that cover a wider frequency range, thus making better use of available space for space-constrained cases. In these instances, miniature antennas with high gain are significantly relevant because the radiating elements are combined into large antenna arrays. Furthermore, metamaterials' negative refractive index focuses electromagnetic radiation by a flat lens versus being dispersed.

The DNG Shell

The earliest research in metamaterial antennas was an analytical study of a miniature dipole antenna surrounded with a metamaterial. This material is known variously as a negative index metamaterial (NIM) or double negative metamaterial (DNG) among other names. This configuration analytically and numerically appears to produce an order of magnitude increase in power. At the same time, the reactance appears to offer a corresponding decrease. Furthermore, the DNG shell becomes a natural impedance matching network for this system.

Ground Plane Applications

Metamaterials employed in the ground planes surrounding antennas offer improved isolation between radio frequency, or microwave channels of (multiple-input multiple-output) (MIMO) antenna arrays. Metamaterial, high-impedance groundplanes can also improve radiation efficiency and axial ratio performance of low-profile antennas located close to the ground plane surface. Metamaterials have also been used to increase beam scanning range by using both the forward and backward waves in leaky wave antennas. Various metamaterial antenna systems can be employed to support surveillance sensors, communication links, navigation systems and command and control systems.

Novel Configurations

Besides antenna miniaturization, the novel configurations have potential applications ranging from radio frequency devices to optical devices. Other combinations, for other devices in metamaterial antenna subsystems are being researched. Either double negative metamaterial slabs are used exclusively or combinations of double positive (DPS) with DNG slabs, or epsilon-negative (ENG) slabs with mu-negative (MNG) slabs are employed in the subsystems. Antenna subsystems that are currently being researched include cavity resonators, waveguides, scatters and antennas (radiators). Metamaterial antennas were commercially available by 2009.

Microwave Lens

The metamaterial lens, found in metamaterial antenna systems, is used as an efficient coupler to external radiation, focusing radiation along or from a microstrip transmission line into transmitting and receiving components. Hence, it can be used as an input device. In addition, it can enhance the amplitude of evanescent waves, as well as correct the phase of propagating waves.

Directing Radiation

In this instance an SRR uses layers of a metallic mesh of thin wires – with wires in the three directions of space and slices of foam. This material's permittivity above the plasma frequency can be positive and less than one. This means that the refractive index is just above zero. The relevant parameter is often the contrast between the permittivities rather than the overall permittivity value at desired frequencies. This occurs because the equivalent (effective) permittivity has a behavior governed by a plasma frequency in the microwave domain. This low optical index material then is a good candidate for extremely convergent microlenses. Methods that have been developed theoretically using dielectric photonic crystals applied in the microwave domain to realize a directive emitter using metallic grids.

In this instance, arrayed wires in a cubic, crystal lattice structure can be analyzed as an array of aerials (antenna array). As a lattice structure it has a lattice constant. The lattice constant or lattice parameter refers to the constant distance between unit cells in a crystal lattice. The earlier discovery of plasmons created the view that metal at plasmon frequency f_p is a composite material. The effect of plasmons on any metal sample is to create properties in the metal such that it can behave as a dielectric, independent of the wave vector of the EM excitation (radiation) field. Furthermore, a minute-fractionally small amount of plasmon energy is absorbed into the system denoted as γ. For aluminium $f_p = 15$ eV, and $\gamma = 0.1$ eV. Perhaps the most important result of the interaction of metal and the plasma frequency is that permittivity is negative below the plasma frequency, all the way to the minute value of γ.

These facts ultimately result in the arrayed wire structure as being effectively a homogeneous medium. This metamaterial allows for control of the direction of emission of an electromagnetic radiation source located inside the material in order to collect all the energy in a small angular domain around the normal. By using a slab of a metamaterial, diverging electromagnetic waves are focused into a narrow cone. Dimensions are small in comparison to the wavelength and thus the slab behaves as a homogeneous material with a low plasma frequency.

Transmission Line Models

Conventional Transmission Lines

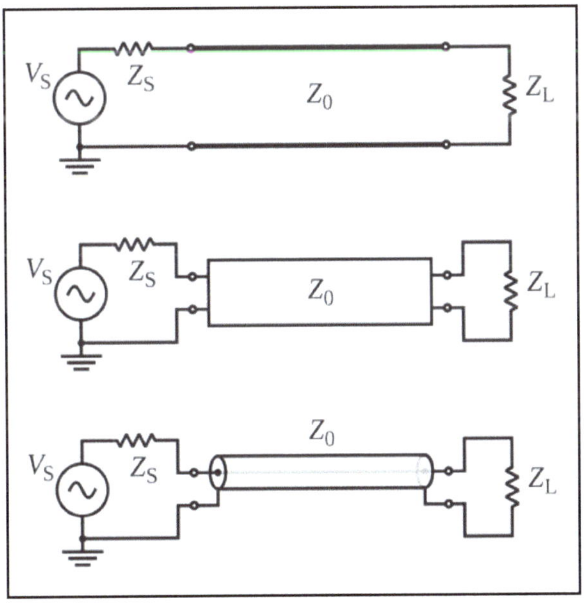

Variations on the Schematic Electronic Symbol for a Transmission Line

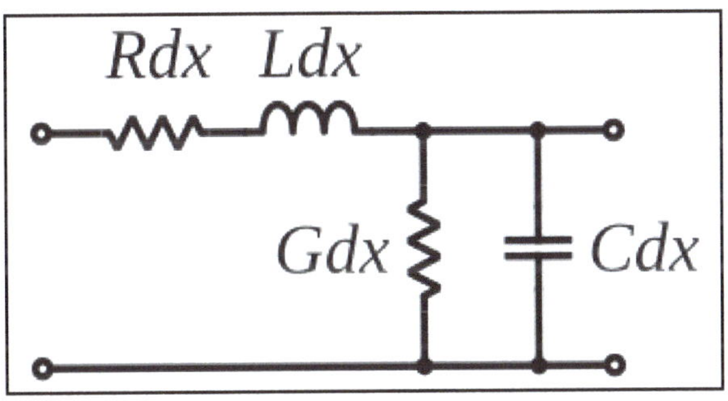

Schematic Representation of the Elementary Components of a Transmission Line

A transmission line is the material medium or structure that forms all or part of a path from one place to another for directing the transmission of energy, such as electromagnetic waves or electric power transmission. Types of transmission line include wires, coaxial cables, dielectric slabs, striplines, optical fibers, electric power lines and waveguides.

A microstrip is a type of transmission line that can be fabricated using printed circuit board technology and is used to convey microwave-frequency signals. It consists of a conducting strip separated from a ground plane by a dielectric layer known as the substrate. Microwave components such as antennas, couplers, filters and power dividers can be formed from a microstrip. From the simplified schematics to the right it can be seen that total impedance, conductance, reactance (capacitance and inductance) and the transmission medium (transmission line) can be represented by single components that give the overall value. With transmission line media it is important to match the load impedance Z_L to the characteristic impedance Z_0 as closely as possible, because it is usually desirable that the load absorbs as much power as possible.

- R is the resistance per unit length,
- L is the inductance per unit length,
- G is the conductance of the dielectric per unit length,
- C is the capacitance per unit length,
- j is the imaginary unit, and
- w is the angular frequency.

Lumped Circuit Elements

Often, because of the goal that moves physical metamaterial inclusions (or cells) to smaller sizes, discussion and implementation of lumped LC circuits or distributed LC networks are often examined. Lumped circuit elements are actually microscopic elements that effectively approximate their larger component counterparts. For example, circuit capacitance and inductance can be created with split rings, which are on the scale of nanometers at optical frequencies. The distributed LC model is related to the lumped LC model, however the distributed-element model is more accurate but more complex than the lumped-element model.

Metamaterial-Loaded Transmission-Line Configurations

Some noted metamaterial antennas employ negative-refractive-index transmission-line metamaterials (NRI-TLM). These include lenses that can overcome the diffraction limit, narrowband and broadband phase-shifting lines, small antennas, low-profile antennas, antenna feed networks, novel power architectures, and high-directivity couplers. Loading a planar metamaterial network of TLs with series capacitors and shunt inductors produces higher performance. This results in a large operating bandwidth while the refractive index is negative.

Because superlenses can overcome the diffraction limit, this allows for a more efficient coupling to external radiation and enables a broader frequency band. For example, the superlens can be applied to the TLM architecture. In conventional lenses, imaging is limited by the diffraction limit. With superlenses the details of the near field images are not lost. Growing evanescent waves are supported in the metamaterial ($n < 1$), which restores the decaying evanescent waves from the source. This results in a diffraction-limited resolution of $\lambda/6$, after some small losses. This compares with $\lambda/2$, the normal diffraction limit for conventional lenses.

By combining right-handed (RHM) with left-handed materials (LHM) as a composite material (CRLH) construction, both a backward to forward scanning capability is obtained. Metamaterials were first used for antenna technology around 2005. This type of antenna used the established capability of SNGs to couple with external radiation. Resonant coupling allowed for a wavelength larger than the antenna. At microwave frequencies this allowed for a smaller antenna.

A metamaterial-loaded transmission line has significant advantages over conventional or standard delay transmission lines. It is more compact in size, it can achieve positive or negative phase shift while occupying the same short physical length and it exhibits a linear, flatter phase response with frequency, leading to shorter group delays. It can work in lower frequency because of high series distributed-capacitors and has smaller plane dimensions than its equivalent coplanar structure.

Negative Refractive Index Metamaterials Supporting 2-D Waves

In 2002, rather than using SRR-wire configuration, or other 3-D media, researchers looked at planar configurations that supported backward wave propagation, thus demonstrating negative refractive index and focusing as a consequence.

It has long been known that transmission lines periodically loaded with capacitive and inductive elements in a high-pass configuration support certain types of backward waves. In addition, planar transmission lines are a natural match for 2-D wave propagation. With lumped circuit elements they retain a compact configuration and can still support the lower RF range. With this in mind, high pass and cutoff, periodically loaded, two-dimensional LC transmission line networks were proposed. The LC networks can be designed to support backward waves, without bulky SRR/wire structure.

This was the first such proposal which veered away from bulk media for a negative refractive effect. A notable property of this type of network is that there is no reliance on resonance, Instead the ability to support backward waves defines negative refraction. The principles behind focusing are derived from Veselago and Pendry. Combining a conventional, flat, (planar) DPS slab, M-1, with a left-handed medium, M-2, a propagating electromagnetic wave with a wave vector k1 in M-1, results in a refracted wave with a wave vector k2 in M-2.

Since, M-2 supports backward wave propagation k2 is refracted to the opposite side of the normal, while the Poynting vector of M-2 is anti-parallel with k2. Under such conditions, power is refracted through an effectively negative angle, which implies an effectively negative index of refraction. Electromagnetic waves from a point source located inside a conventional DPS can be focused inside an LHM using a planar interface of the two media. These conditions can be modeled by exciting a single node inside the DPS and observing the magnitude and phase of the voltages to ground at all points in the LHM. A focusing effect should manifest itself as a "spot" distribution of voltage at a predictable location in the LHM.

Negative refraction and focusing can be accomplished without employing resonances or directly synthesizing the permittivity and permeability. In addition, this media can be practically fabricated by appropriately loading a host transmission line medium. Furthermore, the resulting planar topology permits LHM structures to be readily integrated with conventional planar microwave circuits and devices.

When transverse electromagnetic propagation occurs with a transmission line medium, the analogy for permittivity and permeability is $\varepsilon = L$, and $\mu = C$. This analogy was developed with positive values for these parameters. The next logic step was realizing that negative values could be achieved. In order to synthesize a left-handed medium ($\varepsilon < 0$ and $\mu < 0$) the series reactance and shunt susceptibility should become negative, because the material parameters are directly proportional to these circuit quantities.

A transmission line that has lumped circuit elements that synthesize a left-handed medium is referred to as a "dual transmission line" as compared to "conventional transmission line". The dual transmission line structure can be implemented in practice by loading a host transmission line with lumped series capacitors (C) and shunt inductors (L). In this periodic structure, the loading is strong such that the lumped elements dominate the propagation characteristics.

Left-Handed Behavior in LC Loaded Transmission Lines

Using SRRs at RF frequencies, as with wireless devices, requires the resonators to be scaled to larger dimensions. This worked against making the devices more compact. In contrast, LC network configurations could be scaled to both microwave and RF frequencies. LC-loaded transmission lines enabled a new class of metamaterials to produce a negative refractive index. Relying on LC networks to emulate electrical permittivity and magnetic permeability resulted in a substantial increase in operating bandwidths.

Moreover, their unit cells are connected through a transmission-line network and may be equipped with lumped circuit elements, which permit them to be compact at frequencies where an SRR cannot be compact. The flexibility gained through the use of either discrete or printed elements enables planar metamaterials to be scalable from the megahertz to the tens of gigahertz range. In addition, replacing capacitors with varactors allowed the material properties to be dynamically tuned. The proposed media are planar and inherently support two-dimensional (2-D) wave propagation, making them well-suited for RF/microwave device and circuit applications.

Growing Evanescent Waves in Negative-Refractive-Index Transmission-Line Media

The periodic 2-D LC loaded transmission-line (TL) was shown to exhibit NRI properties over a broad frequency range. This network will be referred to as a dual TL structure since it is of a high-pass configuration, as opposed to the low-pass representation of a conventional TL structure. Dual TL structures have been used to experimentally demonstrate backward-wave radiation and focusing at microwave frequencies. As a negative refractive index medium, a dual TL structure is not simply a phase compensator. It can enhance the amplitude of evanescent waves, as well as correct the phase of propagating waves. Evanescent waves actually grow within the dual TL structure.

Backward Wave Antenna Using an NRI Loaded Transmission Line

Grbic et al. used one-dimensional LC loaded transmission line network, which supports fast backward-wave propagation to demonstrate characteristics analogous to "reversed Cherenkov radiation". Their proposed backward-wave radiating structure was inspired by negative refractive index LC materials. The simulated E-plane pattern at 15 GHz showed radiation towards the backfire direction in the far-field pattern, clearly indicating the excitation of a backward wave. Since the transverse dimension of the array is electrically short, the structure is backed by a long metallic trough. The trough acts as a waveguide below cut-off and recovers the back radiation, resulting in unidirectional far-field patterns.

Planar NIMs with Periodic Loaded Transmission Lines

Planar media can be implemented with an effective negative refractive index. The underlying concept is based on appropriately loading a printed network of transmission lines periodically with inductors and capacitors.

This technique results in effective permittivity and permeability material parameters that are both inherently and simultaneously negative, obviating the need to employ separate means. The proposed media possess other desirable features including very wide bandwidth over which the refractive index remains negative, the ability to guide 2-D TM waves, scalability from RF to millimeter-wave frequencies and low transmission losses, as well as the potential for tunability by inserting varactors and/or switches in the unit cell.

The concept has been verified with circuit and full-wave simulations. A prototype focusing device has been tested experimentally. The experimental results demonstrated focusing of an incident cylindrical wave within an octave bandwidth and over an electrically short area; suggestive of near-field focusing. RF/microwave devices can be implemented based on these proposed media for applications in wireless communications, surveillance and radars.

Larger Transmission Lines

According to some researchers SRR/wire-configured metamaterials are bulky 3-D constructions that are difficult to adapt for RF/microwave device and circuit applications. These structures can achieve a negative index of refraction only within a narrow bandwidth. When applied to wireless devices at RF frequencies the split ring-resonators have to be scaled to larger dimensions, which, in turn forces a larger device size.

The proposed structures go beyond the wire/SRR composites in that they do not rely on SRRs to synthesize the material parameters, thus leading to dramatically increased operating bandwidths. Moreover, their unit cells are connected through a transmission-line network and they may, therefore, be equipped with lumped elements, which permit them to be compact at frequencies where the SRR cannot be compact. The flexibility gained through the use of either discrete or printed elements enables planar metamaterials to be scalable from the megahertz to the tens of gigahertz range. In addition, by utilizing varactors instead of capacitors, the effective material properties can be dynamically tuned. Furthermore, the proposed media are planar and inherently support two-dimensional (2-D) wave propagation. Therefore, these new metamaterials are well suited for RF/microwave device and circuit applications.

In the long-wavelength regime, the permittivity and permeability of conventional materials can be artificially synthesized using periodic LC networks arranged in a low-pass configuration. In the dual (high-pass) configuration, these equivalent material parameters assume simultaneously negative values, and may therefore be used to synthesize a negative refractive index.

Configurations

Antenna theory is based on classical electromagnetic theory as described by Maxwell's equations. Physically, an antenna is an arrangement of one or more conductors, usually called elements. An alternating current is created in the elements by applying a voltage at the antenna terminals, causing the elements to radiate an electromagnetic field. In reception, the reverse occurs: an electromagnetic field from another source induces an alternating current in the elements and a corresponding voltage at the antenna's terminals. Some receiving antennas (such as parabolic and horn types) incorporate shaped reflective surfaces to collect EM waves from free space and direct or focus them onto the actual conductive elements. An antenna creates sufficiently strong electromagnetic fields at large distances. Reciprocally, it is sensitive to the electromagnetic fields impressed upon it externally. The actual coupling between a transmitting and receiving antenna is so small that amplifier circuits are required at both the transmitting and receiving stations. Antennas are usually created by modifying ordinary circuitry into transmission line configurations. The required antenna for any given application is dependent on the bandwidth employed, and range (power) requirements. In the microwave to millimeter-wave range – wavelengths from a few meters to millimeters – the following antennas are usually employed:

Dipole antennas, short antennas, parabolic and other reflector antennas, horn antennas, periscope antennas, helical antennas, spiral antennas, surface-wave and leaky wave antennas. Leaky wave antennas include dielectric and dielectric loaded antennas, and the variety of microstrip antennas.

Radiation Properties with SRRs

The SRR was introduced by Pendry in 1999, and is one of the most common elements of metamaterials. As a nonmagnetic conducting unit, it comprises an array of units that yield an enhanced negative effective magnetic permeability, when the frequency of the incident electromagnetic field is close to the SRR resonance frequency. The resonant frequency of the SRR depends on its shape and physical design. In addition, resonance can occur at wavelengths much larger than its size. For the further shape optimization of the elements it is expedient to use genetic and other optimization algorithms. In multi-frequency designs one may apply fractal designs such as those of Sierpensky, Koch or other fractals instead of SRRs.

Double Negative Metamaterials

Through the application of double negative metamaterials (DNG), the power radiated by electrically small dipole antennas can be notably increased. This could be accomplished by surrounding an antenna with a shell of double negative (DNG) material. When the electric dipole is embedded in a homogeneous DNG medium, the antenna acts inductively rather than capacitively, as it would in free space without the interaction of the DNG material. In addition, the dipole-DNG shell combination increases the real power radiated by more than an order of magnitude over a free space antenna. A notable decrease in the reactance of the dipole antenna corresponds to the increase in radiated power. The reactive power indicates that the DNG shell acts as a natural matching network for the dipole. The DNG material matches the intrinsic reactance of this antenna system to free space, hence the impedance of DNG material matches free space. It provides a natural matching circuit to the antenna.

Single Negative SRR and Monopole Composite

The addition of an SRR-DNG metamaterial increased the radiated power by more than an order of magnitude over a comparable free space antenna. Electrically small antennas, high directivity and tunable operational frequency are produced with negative magnetic permeability. When combining a right-handed material (RHM) with a Veselago-left-handed material (LHM) other novel properties are obtained. A single negative material resonator, obtained with an SRR, can produce an electrically small antenna when operating at microwave frequencies.

The configuration of an SRR assessed was two concentric annular rings with relative opposite gaps in the inner and outer ring. Its geometrical parameters were R = 3.6 mm, r = 2.5 mm, w = 0.2 mm, t = 0.9 mm. R and r are used in annular parameters, w is the spacing between the rings and t = the width of the outer ring. The material had a thickness of 1.6 mm. Permittivity was 3.85 at 4 GHz. The SRR was fabricated with an etching technique onto a 30 µm thick copper substrate.

The SRR was excited by using a monopole antenna. The monopole antenna was composed of a coaxial cable, ground plane and radiating components. The ground plane material was aluminium. The operation frequency of the antenna was 3.52 GHz, which was determined by considering the geometrical parameters of SRR. An 8.32 mm length of wire was placed above the ground plane, connected to the antenna, which was one quarter of the operation wavelength. The antenna worked with a feed wavelength of 3.28 mm and feed frequency of 7.8 GHz. The SRR's resonant frequency was smaller than the monopole operation frequency.

The monopole-SRR antenna operated efficiently at ($\lambda/10$) using the SRR-wire configuration. It demonstrated good coupling efficiency and sufficient radiation efficiency. Its operation was comparable to a conventional antenna at $\lambda/2$, which is a conventional antenna size for efficient coupling and radiation. Therefore, the monopole-SRR antenna becomes an acceptable electrically small antenna at the SRR's resonance frequency.

When the SRR is made part of this configuration, characteristics such as the antenna's radiation pattern are entirely changed in comparison to a conventional monopole antenna. With modifications to the SRR structure the antenna size could reach ($\lambda/40$). Coupling 2, 3, and 4 SRRs side by side slightly shifts radiation patterns.

Patch Antennas

In 2005 a patch antenna with a metamaterial cover was proposed that enhanced directivity. According to the numerical results, the antenna showed significant improvement in directivity, compared to conventional patch antennae. This was cited in 2007 for an efficient design of directive patch antennas in mobile communications using metamaterials. This design was based on the left-handed material (LHM) transmission line model, with the circuit elements L and C of the LHM equivalent circuit model. This study developed formulae to determine the L and C values of the LHM equivalent circuit model for desirable characteristics of directive patch antennas. Design examples derived from actual frequency bands in mobile communications were performed, which illustrates the efficiency of this approach.

Flat Lens Horn Antenna

This configuration uses a flat aperture constructed of zero-index metamaterial. This has advantages over ordinary (conventional) curved lenses, which results in a much improved directivity. These investigations have provided capabilities for the miniaturization of microwave source and non-source devices, circuits, antennas and the improvement of electromagnetic performance.

Metamaterials Surface Antenna Technology

Metamaterials surface antenna technology (M-SAT) is an invention that uses metamaterials to direct and maintain a consistent broadband radio frequency beam locked on to a satellite whether the platform is in motion or stationary. Gimbals and motors are replaced by arrays of metamaterials in a planar configuration. Also, with this new technology phase shifters are not required as with phased array equipment. The desired affect is accomplished by varying the pattern of activated metamaterial elements as needed. The technology is a practical application of metamaterial cloaking theory. The antenna is approximately the size of a laptop computer. Research and applications of metamaterial based antennas. Related components are also researched.

Subwavelength Cavities and Waveguides

When the interface between a pair of materials that function as optical transmission media interact as a result of opposing permittivity and / or permeability values that are either ordinary (positive) or extraordinary (negative), notable anomalous behaviors may occur. The pair would be a DNG metamaterial (layer), paired with a DPS, ENG or MNG layer. Wave propagation behavior and properties may occur that would otherwise not happen if only DNG layers are paired together.

At the interface between two media, the concept of the continuity of the tangential electric and magnetic field components can be applied. If either the permeability or permittivity of two media has opposite signs then the normal components of the tangential field, on both sides of the interface, will be discontinuous at the boundary. This implies a concentrated resonant phenomenon at the interface. This appears to be similar to the current and voltage distributions at the junction between an inductor and capacitor, at the resonance of an L-C circuit. This "interface resonance" is essentially independent of the total thickness of the paired layers, because it occurs along the discontinuity between two such conjugate materials.

Parallel-Plate Waveguiding Structures

The geometry consists of two parallel plates as perfect conductors (PEC), an idealized structure, filled by two stacked planar slabs of homogeneous and isotropic materials with their respective constitutive parameters ε_1, ε_2, u_1, u_2. Each slab has thickness = d, slab 1 = d_1, and slab 2 = d_2. Choosing which combination of parameters to employ involves pairing DPS and DNG or ENG and MNG materials. As mentioned previously, this is one pair of oppositely-signed constitutive parameters, combined.

Thin Subwavelength Cavity Resonators

Phase Compensation:

- The real component values for negative permittivity and permeability results in real component values for negative refraction n. In a lossless medium, all that would exist are real values. This concept can be used to map out phase compensation when a conventional lossless material, DPS, is matched with a lossless NIM (DNG).
- In phase compensation, the DPS of thickness d_1 has $\varepsilon > 0$ and $\mu > 0$. Conversely, the NIM of thickness d_2 has $\varepsilon < 0$ and $\mu < 0$. Assume that the intrinsic impedance of the DPS dielectric material (d_1) is the same as that of the outside region and responding to a normally incident planar wave. The wave travels through the medium without any reflection because the DPS impedance and the outside impedance are equal. However, the plane wave at the end of DPS slab is out of phase with the plane wave at the beginning of the material.
- The plane wave then enters the lossless NIM (d_2). At certain frequencies $\varepsilon < 0$ and $\mu < 0$ and $n < 0$. Like the DPS, the NIM has intrinsic impedance that is equal to the outside, and, therefore, is also lossless. The direction of power flow (i.e., the Poynting vector) in the first slab should be the same as that in the second one, because the power of the incident wave enters the first slab (without any reflection at the first interface), traverses the first slab, exits the second interface, enters the second slab and traverses it, and finally leaves the second slab. However, as stated earlier, the direction of power is anti-parallel to the direction of phase velocity. Therefore, the wave vector k_2 is in the opposite direction of k_1. Furthermore, whatever phase difference is developed by traversing the first slab can be decreased and even cancelled by traversing the second slab. If the ratio of the two thicknesses is $d_1 / d_2 = n_2 / n_1$, then the total phase difference between the front and back faces is zero. This demonstrates how the NIM slab at chosen frequencies acts as a phase compensator. It is important to note that this phase compensation process is only on the ratio of d_1 / d_2 rather than the thickness of $d_1 + d_1$. Therefore, $d_1 + d_1$ can be any value, as long as this ratio satisfies the above condition. Finally, even though this two-layer structure is present, the wave traversing this structure would not experience the phase difference.

- Following this, the next step is the subwavelength cavity resonator.

Compact Subwavelength 1-D Cavity Resonators using Metamaterials:

- The phase compensator described above can be used to conceptualize the possibility of designing a compact 1-D cavity resonator. The above two-layer structure is applied as two perfect reflectors, or in other words, two perfect conducting plates. Conceptually, what is constrained in the resonator is d_1 / d_2, not $d_1 + d_2$. Therefore, in principle, one can have a thin subwavelength cavity resonator for a given frequency, if at this frequency the second layer acts a metamaterial with negative permittivity and permeability and the ratio correlates to the correct values.
- The cavity can conceptually be thin while still resonant, as long as the ratio of thicknesses is satisfied. This can, in principle, provide possibility for subwavelength, thin, compact cavity resonators.

Metamaterial Ground Plane

Leaky Mode Propagation with Metamaterial Ground Plane

A magnetic dipole was placed on metamaterial (slab) ground plane. The metamaterials have either constituent parameters that are both negative, or negative permittivity or negative permeability. The dispersion and radiation properties of leaky waves supported by these metamaterial slabs, respectively, were investigated.

Patented Systems

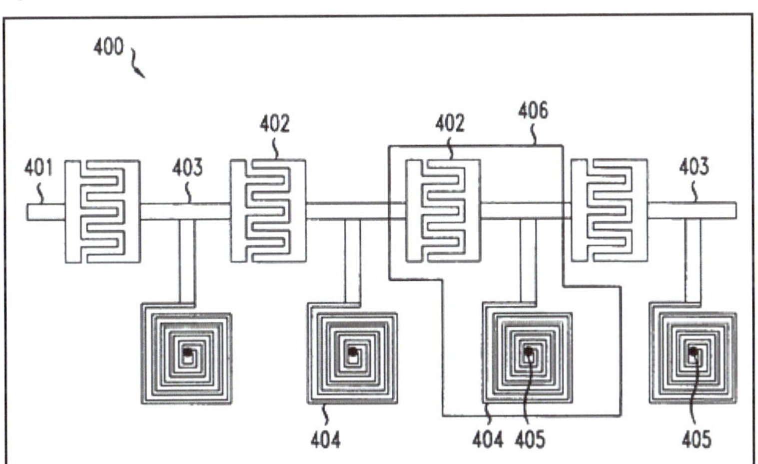

Microstrip Line (400) for a Phased Array Metamaterial Antenna System. 401 Represents Unit-Cell Circuits Composed Periodically along the Microstrip. 402 Series Capacitors. 403 are T-junctions between Capacitors, which Connect (404) Spiral Inductor Delay Lines to 401. 404 are also Connected to Ground vias 405

Phased array systems and antennas for use in such systems are well known in areas such as telecommunications and radar applications. In general phased array systems work by coherently reassembling signals over the entire array by using circuit elements to compensate for relative phase differences and time delays.

Phased Array Antenna

Patented in 2004, one phased array antenna system is useful in automotive radar applications. By using NIMs as a biconcave lens to focus microwaves, the antenna's sidelobes are reduced in size. This equates to a reduction in radiated energy loss, and a relatively wider useful bandwidth. The system is an efficient, dynamically ranged phased array radar system.

In addition, signal amplitude is increased across the microstrip transmission lines by suspending them above the ground plane at a predetermined distance. In other words, they are not in contact with a solid substrate. Dielectric signal loss is reduced significantly, reducing signal attenuation.

This system was designed to boost the performance of the Monolithic microwave integrated circuit (MMIC), among other benefits. A transmission line is created with photolithography. A metamaterial lens, consisting of a thin wire array focuses the transmitted or received signals between the line and the emitter / receiver elements.

The lens also functions as an input device and consists of a number of periodic unit-cells disposed along the line. The lens consists of multiple lines of the same make up; a plurality of periodic unit-cells. The periodic unit-cells are constructed of a plurality of electrical components; capacitors and inductors as components of multiple distributed-element circuits.

The metamaterial incorporates a conducting transmission element, a substrate comprising at least a first ground plane for grounding the transmission element, a plurality of unit-cell circuits composed periodically along the transmission element and at least one via for electrically connecting the transmission element to at least the first ground plane. It also includes a means for suspending this transmission element a predetermined distance from the substrate in a way such that the transmission element is located at a second predetermined distance from the ground plane.

ENG and MNG Waveguides and Scattering Devices

This structure was designed for use in waveguiding or scattering of waves. It employs two adjacent layers. The first layer is an epsilon-negative (ENG) material or a mu-negative (MNG) material. The second layer is either a double-positive (DPS) material or a double-negative (DNG) material. Alternatively, the second layer can be an ENG material when the first layer is an MNG material or the reverse.

3.4.2 Waveguides

A waveguide is a structure that guides waves by restricting the transmission of energy to one direction. Common types of waveguides include acoustic waveguides which direct sound, optical waveguides which direct light, and radio-frequency waveguides which direct electromagnetic waves other than light like radio waves. Without the physical constraint of a waveguide, waves would expand into three-dimensional space and their intensities would decrease according to the inverse square law.

There are different types of waveguides for different types of waves. The original and most common meaning is a hollow conductive metal pipe used to carry high frequency radio waves, particularly microwaves. Dielectric waveguides are used at higher radio frequencies, and transparent dielectric waveguides and optical fibers serve as waveguides for light. In acoustics, air ducts and horns are used as waveguides for sound in musical instruments and loudspeakers, and specially-shaped metal rods conduct ultrasonic waves in ultrasonic machining.

The geometry of a waveguide reflects its function; in addition to more common types that channel the wave in one dimension, there are two-dimensional slab waveguides which confine waves to two dimensions. The frequency of the transmitted wave also dictates the size of a waveguide: each waveguide has a cutoff wavelength determined by its size and will not conduct waves of greater wavelength; an optical fiber that guides light will not transmit microwaves which have a much larger wavelength. Some naturally occurring structures can also act as waveguides. The SOFAR channel layer in the ocean can guide the sound of whale song across enormous distances. Any shape of waveguide can support EM waves, however irregular shapes are difficult to analyse. Commonly used waveguides are rectangular or circular in cross-section.

Uses

The uses of waveguides for transmitting signals were known even before the term was coined. The phenomenon of sound waves guided through a taut wire have been known for a long time, as well as sound through a hollow pipe such as a cave or medical stethoscope. Other uses of waveguides are in transmitting power between the components of a system such as radio, radar or optical devices. Waveguides are the fundamental principle of guided wave testing (GWT), one of the many methods of non-destructive evaluation. Specific examples:

- Optical fibers transmit light and signals for long distances with low attenuation and a wide usable range of wavelengths.
- In a microwave oven a waveguide transfers power from the magnetron, where waves are formed, to the cooking chamber.

- In a radar, a waveguide transfers radio frequency energy to and from the antenna, where the impedance needs to be matched for efficient power transmission.
- Rectangular and circular waveguides are commonly used to connect feeds of parabolic dishes to their electronics, either low-noise receivers or power amplifier/transmitters.
- Waveguides are used in scientific instruments to measure optical, acoustic and elastic properties of materials and objects. The waveguide can be put in contact with the specimen (as in a medical ultrasonography), in which case the waveguide ensures that the power of the testing wave is conserved, or the specimen may be put inside the waveguide (as in a dielectric constant measurement, so that smaller objects can be tested and the accuracy is better.
- A transmission line is a commonly used specific type of waveguide.

Fig: Waveguide Supplying Power for the Argonne National Laboratory

Properties

Propagation Modes and Cutoff Frequencies

A propagation mode in a waveguide is one solution of the wave equations, or, in other words, the form of the wave. Due to the constraints of the boundary conditions, there are only limited frequencies and forms for the wave function which can propagate in the waveguide. The lowest frequency in which a certain mode can propagate is the cutoff frequency of that mode. The mode with the lowest cutoff frequency is the fundamental mode of the waveguide, and its cutoff frequency is the waveguide cutoff frequency.

Propagation modes are computed by solving the Helmholtz equation alongside a set of boundary conditions depending on the geometrical shape and materials bounding the region. The usual assumption for infinitely long uniform waveguides allows us to assume a propagating form for the wave, i.e. stating that every field component has a known dependency on the propagation direction (i.e. z). More specifically, the common approach is to first replace all unknown time-varying fields $u(x,y,z,t)$ (assuming for simplicity to describe the fields in cartesian components) with their complex phasors representation $U(x,y,z)$, sufficient to fully describe any infinitely long single-tone signal at frequency f, (angular frequency $\omega = 2\pi f$), and rewrite the Helmholtz equation and boundary conditions accordingly. Then, every unknown field is forced to have a form like $U(x,y,z) = \hat{U}(x,y)e^{-\gamma z}$, where the γ term represents the propagation constant (still unknown) along the direction along which the waveguide extends to infinity. The Helmholtz equation can be rewritten to accommodate such form and the resulting equality needs to be solved for γ and $\hat{U}(x,y)$, yielding in the end an eigenvalue equation for γ and a corresponding eigenfunction $\hat{U}(x,y)$ for each solution of the former.

The propagation constant γ of the guided wave is complex, in general. For a lossless case, the propagation constant might be found to take on either real or imaginary values, depending on the chosen solution of the eigenvalue equation and on the angular frequency ω. When γ is purely real, the mode is said to be "below cutoff", since the amplitude of the field phasors tends to exponentially decrease with propagation; an imaginary γ, instead, represents modes said to be "in propagation" or "above cutoff", as the complex amplitude of the phasors does not change with z.

Impedance Matching

In circuit theory, the impedance is a generalization of electrical resistance in the case of alternating current, and is measured in ohms (Ω). A waveguide in circuit theory is described by a transmission line having a length and characteristic impedance. In other words, the impedance indicates the ratio of voltage to current of the circuit component (in this case a waveguide) during propagation of the wave. This description of the waveguide was originally intended for alternating current, but is also suitable for electromagnetic and sound waves, once the wave and material properties (such as pressure, density, dielectric constant) are properly converted into electrical terms (current and impedance for example). Impedance matching is important when components of an electric circuit are connected (waveguide to antenna for example): The impedance ratio determines how much of the wave is transmitted forward and how much is reflected. In connecting a waveguide to an antenna a complete transmission is usually required, so an effort is made to match their impedances. The reflection coefficient vcan be calculated using:

$$\Gamma = \frac{Z_2 - Z_1}{Z_2 + Z_1}$$

where,

- Γ (Gamma) is the reflection coefficient (0 denotes full transmission,
- 1 full reflection, and
- 0.5 is a reflection of half the incoming voltage),
- Z_1 and Z_2 are the impedance of the first component (from which the wave enters) and the second component, respectively.

An impedance mismatch creates a reflected wave, which added to the incoming waves creates a standing wave. An impedance mismatch can be also quantified with the standing wave ratio (SWR or VSWR for voltage), which is connected to the impedance ratio and reflection coefficient by:

$$\mathrm{VSWR} = \frac{|V|_{\max}}{|V|_{\min}} = \frac{1+|\Gamma|}{1-|\Gamma|}$$

where,

- $|V|_{\min/\max}$ are the minimum and maximum values of the voltage absolute value, and
- VSWR is the voltage standing wave ratio,
- which value of 1 denotes full transmission, without reflection and thus no standing wave,
- while very large values mean high reflection and standing wave pattern.

Electromagnetic Waveguides

Radio-Frequency Waveguides

Waveguides can be constructed to carry waves over a wide portion of the electromagnetic spectrum, but are especially useful in the microwave and optical frequency ranges. Depending on the frequency, they can be constructed from either conductive or dielectric materials. Waveguides are used for transferring both power and communication signals. In this Military Radar, Microwave Radiation is Transmitted between the Source and the Reflector by a Waveguide The figure suggests that microwaves leave the box in a circularly symmetric mode (allowing the antenna to rotate), then they are converted to a linear mode, and pass through a flexible stage. Their polarisation is then rotated in a twisted stage and finally they irradiate the parabolic antenna.

Optical Waveguides

Waveguides used at optical frequencies are typically dielectric waveguides, structures in which a dielectric material with high permittivity, and thus high index of refraction, is surrounded by a material with lower permittivity. The structure guides optical waves by total internal reflection. An example of an optical waveguide is optical fiber.

Other types of optical waveguide are also used, including photonic-crystal fiber, which guides waves by any of several distinct mechanisms. Guides in the form of a hollow tube with a highly reflective inner surface have also been used as light pipes for illumination applications.

The inner surfaces may be polished metal, or may be covered with a multilayer film that guides light by Bragg reflection (this is a special case of a photonic-crystal fiber). One can also use small prisms around the pipe which reflect light via total internal reflection —such confinement is necessarily imperfect, however, since total internal reflection can never truly guide light within a lower-index core (in the prism case, some light leaks out at the prism corners).

Acoustic Waveguides

An acoustic waveguide is a physical structure for guiding sound waves. Sound in an acoustic waveguide behaves like electromagnetic waves on a transmission line. Waves on a string, like the ones in a tin can telephone, are a simple example of an acoustic waveguide. Another example are pressure waves in the pipes of an organ. The term acoustic waveguide is also used to describe elastic waves guided in micro-scale devices, like those employed in piezoelectric delay lines and in stimulated Brillouin scattering.

Infinite Tubes

A waveguide (or tube) impose a boundary condition on the wave equation such that the wave function must be equal to zero on the boundary and that the allowed region is finite in all but one dimension. An infinitely long cylinder is an example. Mathematically, any tube with a bulge, where the width of the tube increases, admits at least one non-propagating bound state. Using the variational principles, Jeffrey Goldstone and Robert Jaffe have shown a tube of constant width with a twist admits a bound state.

Sound Synthesis

Sound synthesis uses digital delay lines as computational elements to simulate wave propagation in tubes of wind instruments and the vibrating strings of string instruments.

3.5 CLOAKING AND INVISIBILITY APPLICATIONS

Cloaking and invisibility applications exemplify the transformative potential of metamaterials in controlling wave propagation. By engineering permittivity, permeability, and anisotropy, metamaterials guide electromagnetic, acoustic, and elastic waves around objects, reducing scattering and achieving undetectability.

Advances in static, active, and tunable cloaks have demonstrated feasibility across microwave, terahertz, and optical regimes, with applications in defense, sensing, imaging, and medical technologies. While challenges remain in terms of losses, fabrication, and bandwidth, ongoing research continues to expand the capabilities of cloaking systems. The pursuit of invisibility has shifted from science fiction to scientific reality, highlighting the power of metamaterials to reshape the interaction between waves and matter.

The future of cloaking and invisibility applications lies in hybrid and multifunctional metamaterials. Researchers are exploring active cloaks capable of real-time adaptation, multi-frequency cloaks operating across electromagnetic and acoustic spectra, and integrated systems combining sensing, imaging, and invisibility. Advances in nanofabrication, computational design, and material science will enable more practical and scalable cloaking devices. Applications will expand beyond defense and stealth to include biomedical imaging, non-invasive diagnostics, energy-efficient waveguides, and environmental control. The integration of cloaking principles with emerging metamaterial technologies promises transformative innovations in multiple scientific and engineering fields.

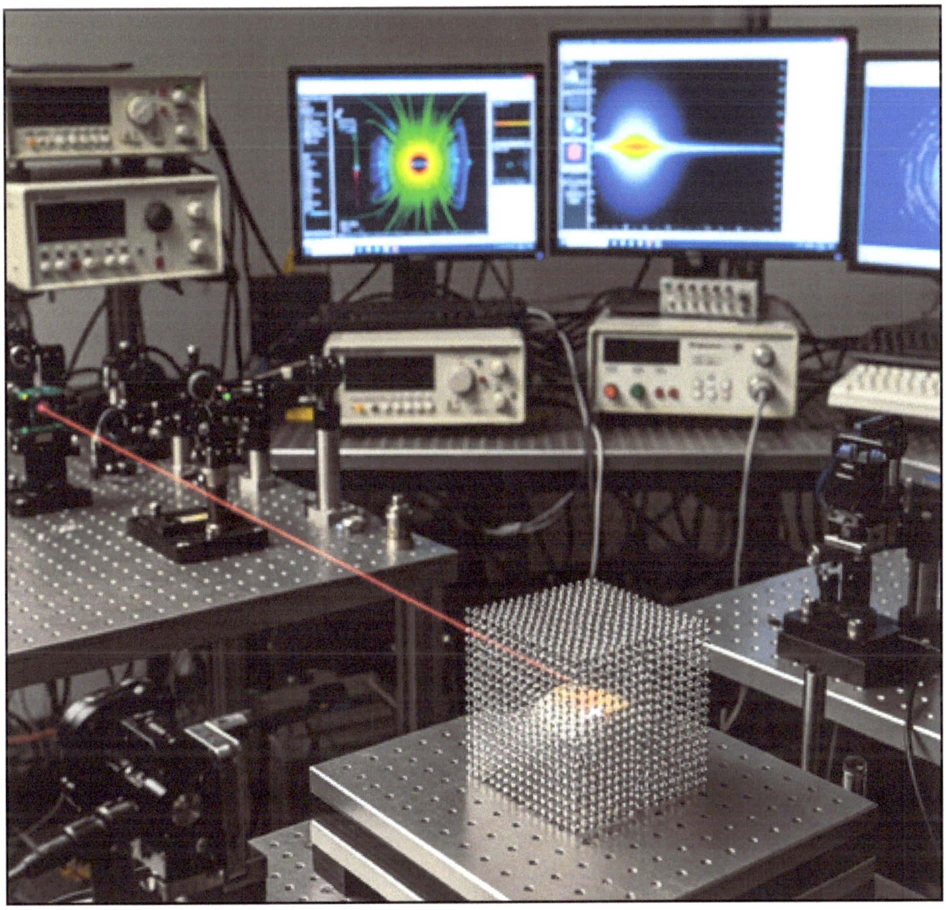

Metamaterial Applications in Cloaking and Invisibility Devices

Despite significant progress, cloaking technologies face substantial challenges. Material losses, particularly at optical frequencies, reduce the effectiveness of cloaks and introduce absorption. Fabrication tolerances for complex, anisotropic, and gradient-index metamaterials are demanding, especially for three-dimensional or optical-scale devices.

Cloaks often operate over narrow frequency ranges, limiting their versatility. Furthermore, achieving omnidirectional cloaking that works for arbitrary wave incidence remains difficult. Ongoing research focuses on low-loss materials, broadband designs, active and tunable systems, and scalable fabrication methods to overcome these limitations and realize practical cloaking devices.

3.5.1 Fundamentals of Cloaking

Cloaking relies on the control of wave trajectories around an object such that waves emerge on the other side as if the object were absent. This requires careful spatial variation of material parameters to redirect incident waves smoothly, eliminating shadows, reflections, or scattering that would reveal the object's presence.

Transformation optics provides the theoretical framework for cloaking, using coordinate transformations to map space and dictate the required permittivity and permeability distributions. The resulting metamaterials are often highly anisotropic and inhomogeneous, designed to mimic the behavior of curved space for electromagnetic waves. Cloaking is fundamentally a wave manipulation problem, requiring both precise material design and subwavelength structural control.

3.5.2 Electromagnetic Cloaking

Electromagnetic cloaking is the most widely studied form of invisibility, where metamaterials manipulate light or other electromagnetic waves to achieve undetectability. Split-ring resonators, wire arrays, and other subwavelength inclusions are arranged to create anisotropic and spatially varying effective permittivity and permeability.

By engineering these responses, the waves bend around the cloaked region, recombining seamlessly on the far side. Early demonstrations of cloaking were achieved in the microwave regime, where fabrication tolerances are less stringent and losses are lower. These experiments provided proof-of-concept results, showing reduced scattering from metallic cylinders and other objects when covered with metamaterial cloaks.

3.5.3 Broadband and Frequency-Selective Cloaks

One challenge in cloaking is achieving broadband performance, as metamaterial resonances are typically narrowband. Frequency-selective cloaks operate effectively over specific ranges, such as microwave or terahertz frequencies, but their invisibility degrades outside the resonant band. To address this, researchers have explored multi-layered designs, hybrid materials, and gradient-index metamaterials that broaden the operational bandwidth. Gradient-index cloaks gradually vary the refractive index, guiding waves around an object over a wider frequency range. While broadband cloaking remains challenging, progress continues in expanding the frequency range, with potential applications in telecommunications, radar stealth, and multi-spectral imaging.

3.5.4 Acoustic and Mechanical Cloaking

Cloaking principles extend beyond electromagnetic waves to acoustic, elastic, and thermal waves. Acoustic cloaks manipulate sound waves, redirecting them around objects to reduce scattering and render them effectively invisible to sonar or audible detection. Similar to electromagnetic cloaks, acoustic cloaks use metamaterials with anisotropic density and bulk modulus to guide sound waves. Mechanical cloaking applies to elastic waves in solids, controlling vibration and stress propagation to shield sensitive components. These extensions demonstrate that the fundamental idea of cloaking—wave redirection via engineered material properties—is versatile, spanning multiple physical domains and offering opportunities for noise reduction, vibration isolation, and structural protection.

3.5.5 Plasmonic and Optical Cloaks

Plasmonic cloaks operate in the optical regime using the interaction between light and metallic nanostructures. By employing thin layers of plasmonic materials with tailored permittivity, it is possible to cancel scattering from an object, achieving partial optical invisibility. Optical cloaks face additional challenges due to material losses and fabrication precision at nanoscales. Despite these difficulties, advances in nanofabrication and metamaterial design have enabled the demonstration of cloaking at visible and near-infrared wavelengths. Plasmonic cloaks are particularly promising for applications in optical sensors, photonic circuits, and stealth technologies, where controlling light at the nanoscale is essential.

3.5.6 Active and Tunable Cloaks

Active cloaking introduces dynamic control over the cloaking effect, allowing the material properties to be adjusted in real time. This can be achieved using tunable metamaterials, liquid crystals, phase-change materials, or electronically controlled components. Active cloaks can adapt to changing wave frequencies, environmental conditions, or object geometries, providing versatility beyond static cloaking devices. Such adaptive systems are valuable for radar evasion, dynamic optical camouflage, and real-time imaging applications. By controlling the response of the cloak actively, it becomes possible to optimize invisibility over broader frequency ranges and varying incident angles.

3.5.7 Applications in Defense and Security

Cloaking has immediate relevance in defense and security applications. Electromagnetic cloaks can reduce radar cross-sections, making vehicles, aircraft, or military assets less detectable. Acoustic cloaks can shield submarines or underwater equipment from sonar detection. Mechanical cloaks can protect sensitive instruments from vibrations or shock waves. Beyond stealth, cloaking technologies can improve sensor networks by minimizing interference and scattering, enhancing situational awareness and detection capabilities. The combination of metamaterials, advanced fabrication, and active control opens new possibilities for security applications across multiple domains.

3.5.8 Applications in Imaging and Sensing

Cloaking also enhances imaging and sensing systems by controlling background scattering and improving signal-to-noise ratios. In medical imaging, metamaterial cloaks can reduce interference from surrounding tissues, improving resolution and contrast. In optical and terahertz systems, cloaks can isolate objects or regions of interest, enabling clearer visualization without distortions caused by surrounding structures. Sensors benefit from cloaking by minimizing scattering and electromagnetic interference, enhancing sensitivity and accuracy. These applications demonstrate that cloaking is not merely about invisibility, but also about manipulating wave propagation to improve measurement and detection performance.

CHAPTER 4

Acoustic and Mechanical Metamaterials

4.1 PHONONIC CRYSTALS AND SOUND MANIPULATION

Phononic crystals provide a versatile and powerful platform for manipulating sound and mechanical waves through engineered periodic structures. By exploiting band gaps, waveguiding, and localized modes, these materials enable noise reduction, vibration control, acoustic imaging, and energy management. Advances in tunable and reconfigurable designs further enhance their potential, allowing adaptive and multifunctional applications. While challenges in fabrication, losses, and broadband operation remain, ongoing research continues to expand the capabilities of phononic crystals. Their ability to control acoustic waves at subwavelength scales demonstrates the transformative potential of metamaterials in sound manipulation, sensing, and energy applications.

The future of phononic crystals lies in multi-functional, tunable, and nanoscale designs that integrate with other metamaterial technologies. Researchers are exploring hybrid acoustic-photonic crystals, reconfigurable lattices, and materials with programmable properties to achieve adaptive control over waves. Applications in energy harvesting, thermal management, medical diagnostics, and noise mitigation will benefit from advances in fabrication, computational optimization, and material science. As understanding of wave-matter interactions continues to grow, phononic crystals are expected to play a central role in next-generation acoustic devices, offering precise control over sound and mechanical energy at multiple scales.

Despite their potential, phononic crystals face challenges in design, fabrication, and material selection. Achieving precise periodicity and structural fidelity is critical to maintain band gap integrity and wave manipulation performance. Material losses, particularly at high frequencies, can reduce the effectiveness of waveguiding and localization. Additionally, creating tunable or broadband phononic crystals requires careful integration of active or responsive materials, which can complicate fabrication and increase system complexity. Addressing these challenges requires advanced fabrication techniques, computational modeling, and material innovations to realize practical and scalable phononic crystal devices.

4.1.1 Structure and Design Principles

Phononic crystals are composed of periodic arrangements of inclusions embedded in a host material, forming a lattice that affects wave propagation. The inclusions can be solid, fluid, or gas-filled, and their size, shape, and material properties determine the crystal's band structure. By adjusting these parameters, engineers can tailor the band gaps to specific frequencies, allowing selective blocking or guiding of sound waves. Two-dimensional and three-dimensional phononic crystals offer enhanced control over acoustic waves, while one-dimensional structures such as layered media are used for simpler filtering applications. The design of phononic crystals relies on the interplay between periodicity, material contrast, and structural geometry to achieve the desired wave manipulation effects.

4.1.2 Band Gaps and Wave Control

A key feature of phononic crystals is the formation of band gaps, frequency ranges in which acoustic waves cannot propagate through the crystal. These band gaps arise from Bragg scattering, where waves interfere destructively due to the periodicity of the structure, or from local resonance effects in the inclusions. The presence of band gaps allows phononic crystals to act as acoustic insulators, filters, or waveguides, depending on the intended application. By carefully designing the lattice parameters and material contrasts, it is possible to create multiple band gaps, widen the forbidden frequency ranges, or even achieve tunable band gaps for adaptive acoustic control.

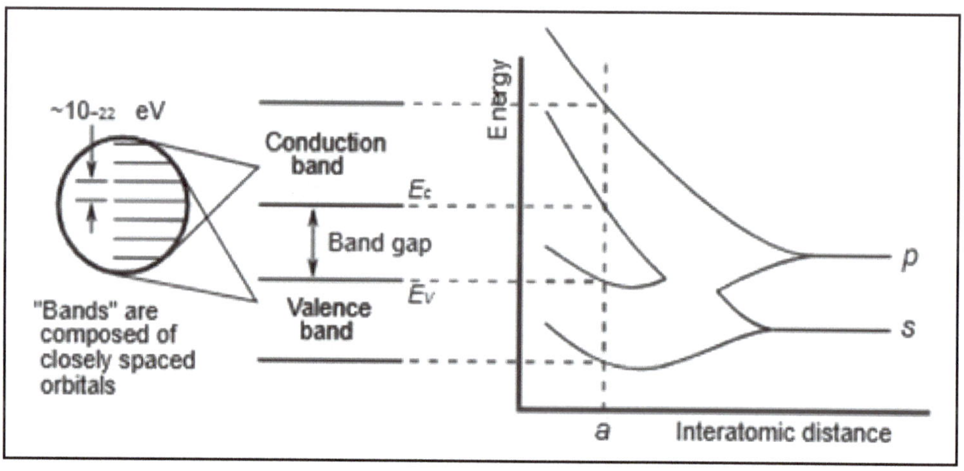

Band Gap

4.1.3 Acoustic Waveguiding and Localization

Phononic crystals enable the guiding and localization of sound waves with high precision. Defects or variations in the periodic lattice can create waveguides that channel acoustic energy along predetermined paths without significant loss. Similarly, localized modes can be engineered around defects or inclusions, concentrating energy in small regions for enhanced interactions. These capabilities are essential for applications in acoustic circuits, ultrasonic sensors, and energy harvesting devices, where precise control over wave propagation and localization is required. Waveguiding and localization in phononic crystals mimic the behavior of electrons in semiconductors and light in photonic crystals, providing a versatile platform for controlling mechanical waves.

4.1.4 Noise Reduction and Vibration Control

One of the most practical applications of phononic crystals is noise reduction and vibration control. By designing band gaps that coincide with the frequencies of unwanted vibrations or sound, phononic crystals can block, reflect, or redirect these waves, creating acoustic shields or vibration isolators. This approach is highly effective in industrial environments, transportation systems, and building acoustics, where controlling noise and vibrations is critical for safety and comfort. The tunability and scalability of phononic crystals allow them to address a wide range of frequencies, from low-frequency structural vibrations to high-frequency ultrasonic noise.

4.1.5 Acoustic Lenses and Imaging

Phononic crystals can manipulate sound waves in ways that enable subwavelength imaging and acoustic focusing. By controlling the phase velocity and refraction of waves within the crystal, engineers can design acoustic lenses that focus sound beyond the diffraction limit. These capabilities are particularly useful in ultrasonic imaging, where improved resolution is essential for medical diagnostics, non-destructive testing, and material characterization. Phononic crystals also enable acoustic cloaking, where waves are guided around objects to reduce scattering and improve imaging performance. The combination of band gap engineering, waveguiding, and lensing makes phononic crystals a powerful tool for advanced acoustic applications.

4.1.6 Tunable and Reconfigurable Phononic Crystals

Recent advances have focused on tunable and reconfigurable phononic crystals, which allow dynamic control over band gaps, waveguiding, and localization. By incorporating materials with variable elastic properties, temperature-dependent behavior, or active elements such as piezoelectric actuators, phononic crystals can adapt their acoustic response in real time. This tunability enables adaptive noise reduction systems, reconfigurable acoustic filters, and active vibration control devices. Reconfigurable phononic crystals expand the functionality of traditional designs, allowing devices to respond to changing environmental conditions or operational requirements.

4.1.7 Thermal and Mechanical Wave Control

Beyond acoustic waves, phononic crystals can also influence thermal and mechanical wave propagation. By affecting phonon transport, which governs heat conduction in solids, phononic crystals can act as thermal insulators or guides, controlling the flow of energy at the nanoscale. Similarly, elastic waves in mechanical structures can be manipulated to concentrate or dissipate energy, providing vibration damping or mechanical energy harvesting capabilities. These effects demonstrate the versatility of phononic crystals, allowing control over multiple forms of wave-based energy in a single platform.

4.1.8 Applications in Ultrasonics and Sensing

Phononic crystals have significant potential in ultrasonic applications and sensing technologies. They can enhance signal resolution, reduce noise, and improve sensitivity in ultrasonic imaging systems. By guiding and localizing acoustic waves, phononic crystals enable precise detection of defects, structural changes, or biological features. Additionally, sensors based on phononic crystals can detect variations in pressure, density, or material composition by observing shifts in resonance frequencies or band gaps. These applications highlight the role of phononic crystals in improving measurement accuracy and enabling new sensing capabilities.

4.2 NEGATIVE MASS DENSITY AND BULK MODULUS

Negative mass density and bulk modulus represent groundbreaking advancements in metamaterials, enabling control over mechanical waves in ways impossible with natural materials. Through resonant structures and carefully engineered composites, these metamaterials exhibit frequency-dependent behaviors that allow backward wave propagation, acoustic cloaking, subwavelength focusing, and vibration isolation. While challenges remain in losses, fabrication, and bandwidth, ongoing research is expanding their capabilities and applications across acoustics, mechanics, and energy systems. These concepts demonstrate the transformative potential of metamaterials in wave manipulation, opening new frontiers in science and technology.

Despite their promise, metamaterials with negative mass density and bulk modulus face several challenges. Losses in the resonant elements can reduce efficiency, limit the strength of negative parameters, and narrow the operational bandwidth. Fabrication at small scales, particularly for high-frequency acoustic or ultrasonic applications, requires precision and consistency in the geometry of resonators. Achieving broadband operation is difficult due to the narrow resonance ranges, though multi-resonator and graded structures can help extend performance. Additionally, integrating these materials into practical devices without introducing unwanted reflections or impedance mismatches remains a technical challenge. Future research in negative mass density and bulk modulus focuses on tunable, reconfigurable, and multi-functional metamaterials. Incorporating active elements, such as piezoelectric actuators, allows dynamic control of the resonant frequency and effective parameters. Nanoscale designs aim to manipulate ultrasonic and even terahertz mechanical waves for biomedical imaging, energy harvesting, and microelectromechanical systems. Hybrid metamaterials combining acoustic and optical properties offer possibilities for multi-physical wave control. Advances in computational design, additive manufacturing, and material science will enable more robust, scalable, and versatile metamaterials capable of unprecedented control over sound and mechanical waves.

4.2.1 Fundamentals of Mass Density and Bulk Modulus

Mass density and bulk modulus are fundamental material parameters governing the propagation of sound and elastic waves. Mass density determines the inertia of a material in response to an applied force, while bulk modulus measures the material's resistance to uniform compression.

In conventional materials, both parameters are positive, meaning that applied forces produce motion and compression in the expected direction. Negative mass density implies that the material accelerates in a direction opposite to the applied force, while negative bulk modulus indicates that the material expands under compression. These counterintuitive behaviors cannot occur in natural materials but can be engineered in metamaterials through resonant structures, periodic inclusions, and subwavelength geometries.

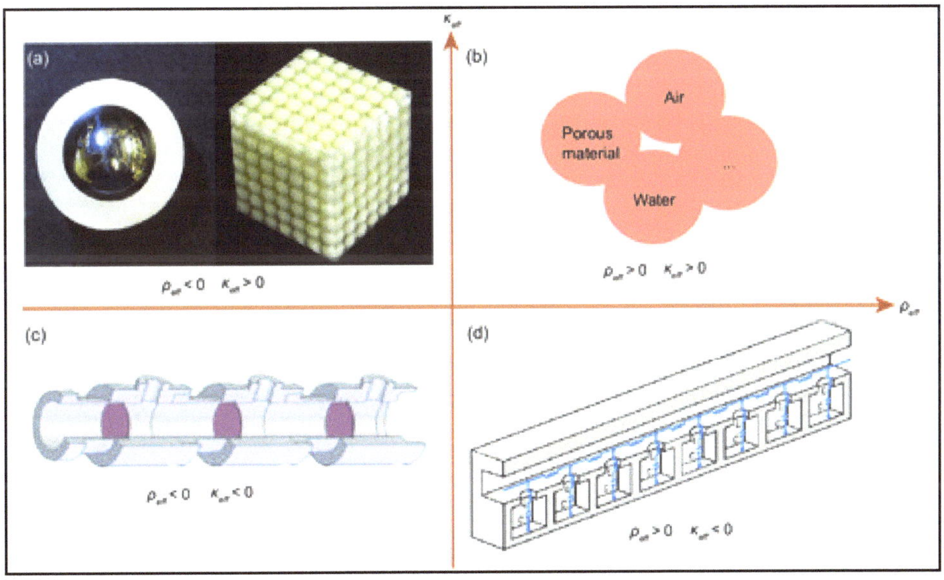

The Parameter Space for the Mass Density and Bulk Modulus

4.2.2 Achieving Negative Mass Density

Negative mass density is typically achieved using locally resonant structures that respond dynamically to incident waves. These resonators, which may include masses connected by springs, membranes, or inertial elements, oscillate out of phase with the incident wave. When the frequency of the wave matches the resonance frequency of the inclusions, the collective response of the system produces an effective negative mass density.

This behavior is frequency-dependent and occurs over a limited range, where the inertia of the resonant elements dominates the overall motion. By tuning the geometry, mass, and stiffness of the inclusions, engineers can control the frequency range and magnitude of negative mass density, enabling precise manipulation of wave propagation and energy localization.

4.2.3 Negative Bulk Modulus and Acoustic Response

Negative bulk modulus arises from resonant structures that respond anomalously to compressive forces. Helmholtz resonators, spherical cavities, or layered composites can produce effective negative bulk modulus when the resonators expand in response to compression at specific frequencies. This effect enables materials to support backward-propagating waves, suppress certain frequency components, or create acoustic band gaps where wave propagation is forbidden. The combination of negative mass density and negative bulk modulus allows for complete control over the acoustic refractive index, leading to exotic phenomena such as negative refraction, acoustic superlensing, and cloaking.

4.2.4 Coupled Resonances and Metamaterial Design

The realization of negative mass density and bulk modulus relies on coupling between resonant elements and the host material. In composite metamaterials, resonators embedded in a matrix interact with incident waves to produce effective material parameters that differ from the constituent materials. These interactions can be tailored to achieve desired frequency-dependent responses, such as simultaneous negative mass density and negative bulk modulus, which are necessary for double-negative acoustic metamaterials. Design considerations include the size, shape, and distribution of resonators, the mechanical properties of the matrix, and the intended frequency range. Computational modeling and finite element analysis are often used to predict the behavior of complex metamaterial structures before fabrication.

4.2.5 Frequency-Dependent Behavior

Negative mass density and bulk modulus are inherently frequency-dependent phenomena. At frequencies below the resonance, the metamaterial behaves like a conventional positive-density and positive-modulus material. Near the resonance, the effective parameters become negative, leading to unusual wave propagation characteristics such as backward waves or suppressed transmission.

Above the resonance, the parameters return to positive values, restoring conventional behavior. This tunable frequency response allows engineers to design metamaterials that target specific acoustic or vibrational frequencies for applications such as noise reduction, vibration isolation, and subwavelength waveguiding. The narrow frequency ranges associated with resonances present challenges for broadband applications, which can be addressed through multi-resonator designs or gradient structures.

4.2.6 Applications in Wave Manipulation

Negative mass density and bulk modulus have enabled a wide range of wave manipulation applications. One notable application is acoustic cloaking, where waves are guided around an object without scattering, effectively rendering it invisible to sound. Double-negative metamaterials with negative mass density and bulk modulus allow acoustic waves to bend backward, achieving cloaking and focusing effects.

These properties also enable subwavelength acoustic lenses, which overcome the diffraction limit and produce high-resolution imaging. Vibration isolation is another important application, where negative mass density structures can cancel or redirect mechanical waves, protecting sensitive equipment from external vibrations. These capabilities demonstrate the transformative potential of double-negative metamaterials in controlling mechanical and acoustic energy.

4.2.7 Experimental Realizations

Experimental demonstrations of negative mass density and bulk modulus have been achieved using various resonant structures. Membrane-type acoustic metamaterials, consisting of tensioned membranes with attached masses, exhibit negative mass density in targeted frequency bands. Helmholtz resonator arrays and layered composites have been used to produce negative bulk modulus, while combined designs achieve double-negative behavior. Measurements typically involve evaluating transmission, reflection, and phase velocity of acoustic waves through the metamaterial. The experimental results often match theoretical predictions, confirming the effectiveness of these engineered structures in producing counterintuitive wave responses.

4.3 VIBRATION ISOLATION AND DAMPING SYSTEMS

Vibration isolation and damping systems are essential for controlling mechanical energy in structures and machinery, protecting equipment, enhancing performance, and improving safety and comfort. Traditional solutions such as springs, elastomers, and tuned mass dampers are complemented by metamaterials and adaptive technologies, offering unprecedented control over vibration amplitude, frequency, and direction. Applications span structural engineering, aerospace, transportation, industrial machinery, and precision equipment, demonstrating the broad impact of these technologies. Continued advances in materials, design, and control strategies will enable more efficient, tunable, and multifunctional systems, shaping the future of vibration management and mechanical wave control.

Despite significant progress, vibration isolation and damping systems face several challenges. Designing systems for broadband vibration suppression without compromising structural stability can be complex. Material properties, especially at high frequencies, may limit damping effectiveness. Active systems require power, sensing, and control infrastructure, adding complexity and potential failure points. Fabrication of metamaterial-based solutions can be challenging and costly, particularly for large-scale applications. Addressing these limitations involves optimizing system design, exploring novel materials, and integrating passive, active, and metamaterial approaches for robust, efficient vibration control.

Future research in vibration isolation and damping focuses on multifunctional, tunable, and lightweight solutions. Advances in metamaterials, smart materials, and additive manufacturing will enable compact systems with tailored frequency responses and adaptive capabilities. Integration with sensors and real-time control systems will provide autonomous adjustment to changing conditions, improving performance and energy efficiency. Applications will expand into robotics, renewable energy, microelectronics, and healthcare, where precise vibration management is increasingly critical. The combination of traditional engineering principles with advanced material design promises transformative improvements in vibration control technologies.

4.3.1 Fundamentals of Vibration Isolation

Vibration isolation involves preventing the transmission of mechanical vibrations from a source to a target structure. This is typically achieved by interposing materials or systems that attenuate energy transfer. Springs, rubber mounts, and air cushions are common traditional isolators that provide compliance and absorb kinetic energy. The performance of these systems depends on parameters such as natural frequency, stiffness, damping ratio, and mass distribution. Effective isolation occurs when the isolator's natural frequency is lower than the frequency of the disturbing vibrations, allowing the isolator to decouple the target from the source. Proper design ensures minimal transmission of vibrational energy, enhancing the performance and longevity of machinery or sensitive instruments.

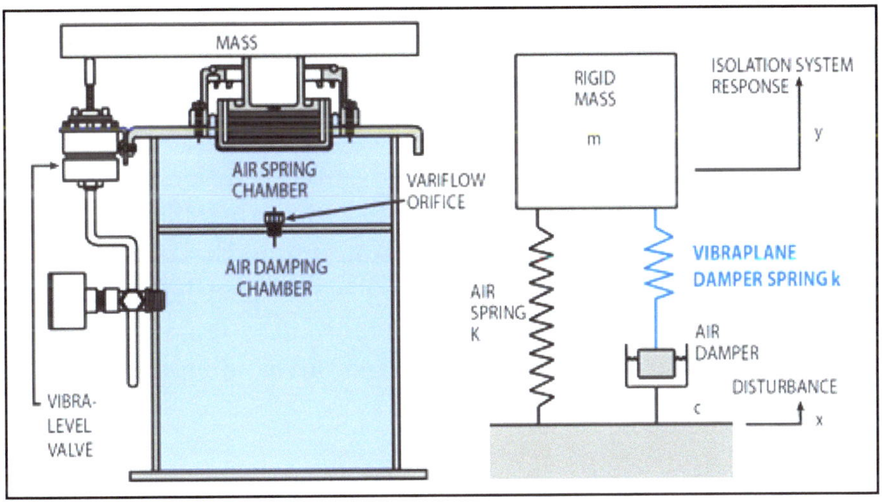

Vibration Isolation

4.3.2 Principles of Damping

Damping systems dissipate mechanical energy, converting vibrational energy into heat or other forms of energy, thereby reducing oscillation amplitude. Materials with inherent viscoelastic properties, such as polymers and composites, provide passive damping by resisting deformation. Additional methods include tuned mass dampers, which use secondary oscillating masses to counteract vibrations at specific frequencies, and viscous fluid dampers that dissipate energy through shear in a fluid medium. Damping can be frequency-dependent, and its effectiveness is enhanced by tuning the damping properties to the predominant vibration modes. High damping ratios reduce resonant amplification, prevent structural fatigue, and improve comfort and performance in mechanical and civil engineering systems.

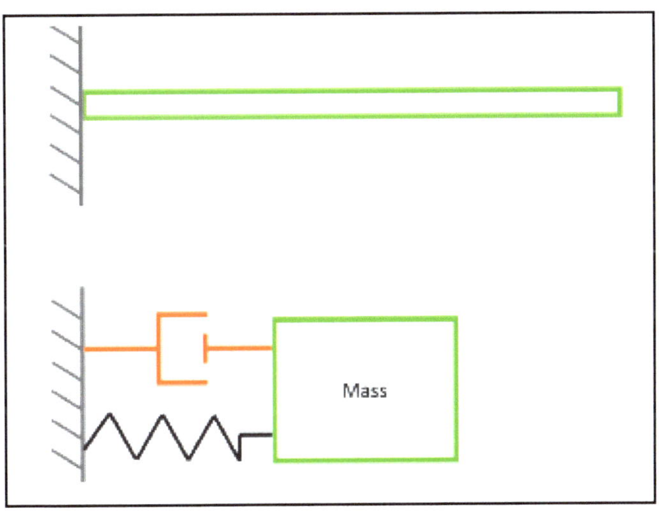

The Principles of Damping

4.3.3 Vibration Isolation in Mechanical Systems

Mechanical systems such as engines, turbines, and industrial machinery generate vibrations due to rotating or reciprocating components. Isolation in these systems is achieved through mounts, resilient pads, or floating foundations that decouple the machinery from the supporting structure. Advanced isolation solutions integrate elastomeric materials, springs, and dampers in series or parallel configurations to achieve tailored vibration attenuation. By carefully tuning the stiffness and damping properties, engineers can target specific vibration frequencies while maintaining structural stability. This approach reduces transmitted forces, minimizes noise, and prevents damage to both the machinery and surrounding structures, ensuring reliable and efficient operation.

4.3.4 Damping in Structural Engineering

Vibration damping is particularly important in structural engineering, where buildings, bridges, and towers are subject to dynamic loads such as wind, earthquakes, and traffic. Tuned mass dampers are commonly used in tall buildings to counteract sway by introducing a secondary mass that oscillates out of phase with the primary structure. Viscous dampers and friction-based devices convert vibrational energy into heat, reducing motion and structural stress.

The design of damping systems in civil structures considers factors such as natural frequencies, modal shapes, and expected load spectra. Effective damping enhances structural resilience, occupant comfort, and longevity, while minimizing maintenance requirements and preventing catastrophic failures during extreme events.

4.3.5 Acoustic and Vibration Coupling

Vibration isolation and damping also play a critical role in controlling acoustic noise. Vibrations in machinery or structural elements can generate airborne sound, leading to noise pollution and reduced operational efficiency. By isolating sources of vibration and dissipating mechanical energy, damping systems reduce sound radiation, improving acoustic performance. In high-precision environments such as laboratories, hospitals, and data centers, vibration-induced noise can interfere with sensitive measurements or equipment operation. Integrating vibration isolation with acoustic design principles ensures both mechanical stability and acoustic control, enhancing overall system performance.

4.3.6 Metamaterial Approaches to Vibration Control

Recent advances in metamaterials have introduced new possibilities for vibration isolation and damping. Acoustic and elastic metamaterials, designed with periodic microstructures, can exhibit negative mass density, negative modulus, or band gaps where wave propagation is forbidden. These properties allow precise control over vibration transmission, enabling frequency-selective isolation and subwavelength attenuation. Locally resonant structures embedded in metamaterials can absorb or redirect energy at targeted frequencies, providing tunable and adaptive damping solutions. By leveraging metamaterial design principles, engineers can create compact, lightweight, and highly effective isolation systems that outperform traditional approaches.

4.3.7 Tunable and Adaptive Systems

Modern vibration isolation and damping systems increasingly incorporate tunable and adaptive features. Active systems use sensors and actuators to monitor vibrations and apply counteracting forces in real time, achieving dynamic isolation and damping. Piezoelectric, electromagnetic, and magnetorheological devices provide precise control over stiffness, damping, and energy dissipation. Adaptive systems are particularly valuable in environments with varying loads or frequencies, such as aerospace vehicles, high-speed trains, and precision manufacturing equipment. By adjusting properties in real time, these systems optimize vibration control, reduce energy consumption, and enhance performance under changing operating conditions.

4.3.8 Applications in Transportation and Aerospace

Vibration isolation and damping are critical in transportation and aerospace applications, where mechanical vibrations can compromise safety, comfort, and structural integrity. In automobiles, suspension systems combine springs, dampers, and resilient mounts to isolate passengers from road-induced vibrations. Aircraft and spacecraft employ vibration control systems to protect delicate avionics, payloads, and crew from aerodynamic and engine-induced vibrations. Tuned mass dampers and active vibration control devices are used to minimize resonant amplification, reduce fatigue, and ensure precise operation of onboard systems. Advanced metamaterial-based solutions offer opportunities to achieve lightweight, high-performance isolation and damping in these demanding environments.

4.3.9 Industrial Machinery and Precision Equipment

Precision equipment in manufacturing, laboratory research, and medical diagnostics requires vibration-free environments to maintain accuracy and functionality. Semiconductor fabrication tools, electron microscopes, and laser systems are highly sensitive to even minute vibrations. Isolation tables, active damping platforms, and metamaterial-based vibration absorbers are employed to protect such equipment from environmental disturbances. By controlling the amplitude, frequency, and direction of mechanical waves, these systems ensure repeatable performance, high-resolution imaging, and process reliability.

Industrial machinery also benefits from reduced maintenance and extended component lifetimes due to minimized vibrational stress.

4.4 TOPOLOGICAL MECHANICAL METAMATERIALS

Topological mechanical metamaterials represent a revolutionary approach to controlling mechanical waves, vibrations, and structural deformations. By exploiting lattice geometry, symmetry-breaking, and topological protection, these materials support robust edge-localized modes, tunable band gaps, and programmable mechanical behavior. Applications span vibration isolation, waveguiding, acoustic control, adaptive structures, and energy transport, demonstrating their versatility and transformative potential. While challenges in fabrication, bandwidth, and integration remain, ongoing research continues to push the boundaries of design, scalability, and functionality. Topological mechanical metamaterials exemplify the power of combining geometry, physics, and materials science to achieve unprecedented control over mechanical systems.

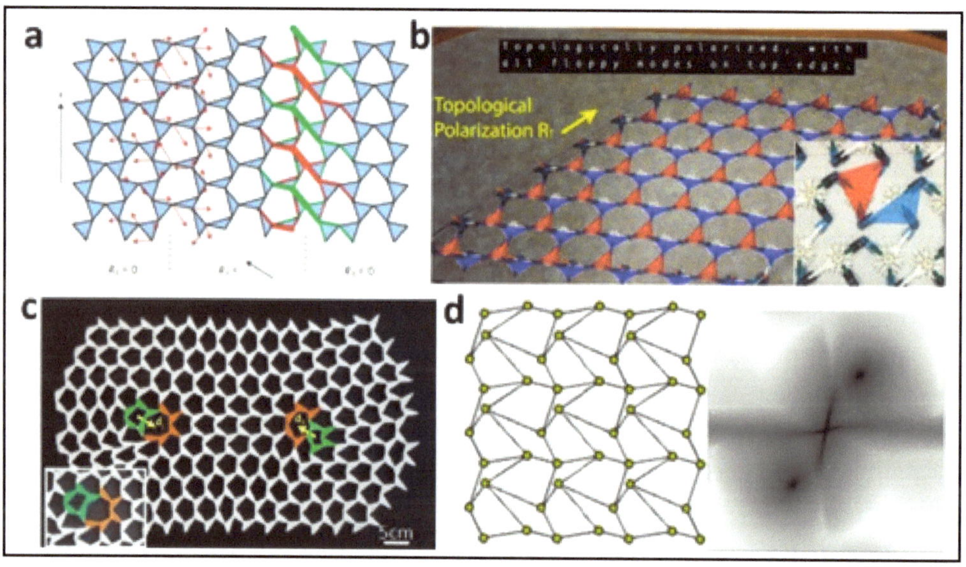

Metamaterials with Topologically Protected Features

Despite their transformative potential, topological mechanical metamaterials face several challenges. Fabrication of precise, small-scale lattice structures can be difficult and costly, particularly for three-dimensional or micro-structured designs. Band gaps and topological properties are often limited to narrow frequency ranges, requiring careful tuning for specific applications.

Integration with conventional mechanical systems may introduce compatibility issues or additional constraints. Furthermore, dynamic or programmable designs necessitate active control systems, increasing complexity and potential points of failure. Addressing these challenges requires advances in fabrication technologies, materials, and computational design methods. Future research in topological mechanical metamaterials focuses on expanding their functionality, tunability, and scalability. Hybrid systems combining active and passive elements, programmable lattices, and multi-physical interactions (acoustic, mechanical, thermal) are under development. Applications in soft robotics, deployable structures, vibration control, energy harvesting, and adaptive mechanical systems are expected to grow. Advances in additive manufacturing, 3D printing, and nanoscale fabrication will enable more complex and efficient designs. As the understanding of topological mechanics deepens, these metamaterials are likely to transform engineering approaches to wave propagation, vibration management, and mechanical design.

4.4.1 Fundamentals of Topological Mechanics

The field of topological mechanics is inspired by topological insulators in condensed matter physics, which conduct electrons along their surfaces while remaining insulating in the bulk. Mechanical analogs achieve similar phenomena for vibrational or elastic waves. By designing the lattice geometry and connectivity of a mechanical metamaterial, it is possible to create topologically protected edge modes that localize at boundaries or interfaces. These edge modes propagate without backscattering, even in the presence of defects or disorder. The robustness of these modes arises from the topological invariants of the system, mathematical quantities that remain unchanged under continuous deformation, ensuring predictable and stable mechanical behavior.

4.4.2 Lattice Design and Geometric Considerations

Topological mechanical metamaterials are often composed of periodic lattices with carefully designed unit cells. Common structures include kagome, honeycomb, and gyroid lattices, which allow the engineering of mechanical band structures and the introduction of band gaps for vibrational waves. By breaking certain symmetries, introducing chirality, or incorporating rotational degrees of freedom, engineers can induce nontrivial topological phases that support edge-localized or defect-immune modes. The design process involves both theoretical modeling and computational simulations to predict band structures, edge states, and response under mechanical loading. Geometric tunability provides a versatile platform for tailoring mechanical properties and wave behavior across different scales.

4.4.3 Topologically Protected Edge Modes

A defining feature of topological mechanical metamaterials is the existence of edge modes that are robust against defects and disorder. These modes are localized at the boundaries of the lattice and can propagate without reflection or scattering, even when encountering irregularities. This property contrasts sharply with conventional mechanical waves, which are highly sensitive to imperfections.

Edge-localized vibrations in topological metamaterials enable highly efficient waveguiding, energy transport, and vibration control. Their robustness ensures reliable performance in practical applications, making topological mechanical metamaterials ideal for environments where conventional waveguides would fail due to imperfections or structural variability.

4.4.4 Band Gaps and Wave Manipulation

Topological mechanical metamaterials can support both bulk band gaps and topologically protected edge states. Band gaps prevent the propagation of waves within specific frequency ranges, while edge states allow selective propagation along interfaces or boundaries. By carefully designing the lattice parameters and symmetry-breaking elements, engineers can control the location and width of band gaps and the frequency range of edge modes. This capability enables precise wave manipulation, including filtering, wave steering, vibration isolation, and directional energy transport. The combination of band gap engineering and topological protection provides a versatile toolkit for designing mechanical systems with tailored wave propagation characteristics.

4.4.5 Vibration Isolation and Energy Transport

The unique properties of topological mechanical metamaterials make them highly effective for vibration isolation and energy transport. Edge-localized modes can guide vibrational energy along predetermined paths without loss or scattering, allowing precise delivery or dissipation of mechanical energy. These materials can isolate sensitive components from unwanted vibrations while directing energy to regions where it can be absorbed or utilized. Applications include protection of delicate instruments, enhancement of mechanical sensors, and efficient routing of mechanical energy in robotics or microelectromechanical systems. The ability to control vibrational energy with topological robustness ensures high reliability and performance under varying environmental conditions.

4.4.6 Programmable and Reconfigurable Systems

Topological mechanical metamaterials can be designed to be programmable or reconfigurable, allowing dynamic control over mechanical properties and wave propagation. By adjusting lattice geometry, connectivity, or local stiffness, the topological phase of the material can be altered, switching edge modes on or off or redirecting vibrational energy along different paths. Reconfigurable designs may incorporate soft materials, hinges, or active components such as actuators to achieve tunability. This adaptability enables the creation of multifunctional structures capable of responding to environmental changes, operational demands, or targeted mechanical tasks. Programmable topological metamaterials hold promise for robotics, adaptive architecture, and deployable structures.

4.4.7 Applications in Mechanical Waveguiding

One of the most significant applications of topological mechanical metamaterials is mechanical waveguiding. Conventional waveguides suffer from scattering and reflection at defects or sharp bends, reducing efficiency. In contrast, topologically protected waveguides allow mechanical energy to travel along edges or interfaces with minimal loss, even in complex geometries or defective lattices. This property is valuable for microelectromechanical systems, precision machinery, and vibration control in civil structures. By guiding energy along predetermined paths, topological metamaterials enable targeted mechanical signal transmission, enhancing both efficiency and reliability.

4.4.8 Acoustic and Elastic Metamaterial Applications

Topological mechanical metamaterials extend beyond structural mechanics to acoustic and elastic wave control. They can manipulate sound waves, elastic vibrations, and surface acoustic waves in a controlled manner. Applications include acoustic waveguides, vibration suppression in buildings, energy harvesting, and high-fidelity sensors. By combining topological protection with conventional metamaterial strategies, such as resonant inclusions or locally varying stiffness, engineers can achieve precise control over wave speed, amplitude, and direction. These capabilities open avenues for innovative devices in ultrasonics, sonar, medical imaging, and noise reduction.

4.5 APPLICATIONS IN STRUCTURAL HEALTH MONITORING

Structural Health Monitoring is a transformative technology that enhances safety, performance, and lifespan of engineering structures. By combining advanced sensors, wave-based techniques, smart materials, and data-driven analytics, SHM enables real-time assessment of structural integrity and early detection of damage. Applications span civil infrastructure, aerospace, mechanical systems, and industrial machinery, providing critical insights for maintenance and operational decisions. While challenges in sensor integration, data processing, and environmental effects remain, ongoing research in smart materials, metamaterials, and machine learning continues to advance the capabilities of SHM systems. These innovations ensure more reliable, efficient, and adaptive monitoring strategies, ultimately safeguarding structures and systems across diverse engineering domains.

Despite its potential, SHM faces several challenges. Sensor placement, calibration, and durability affect measurement accuracy and reliability. Environmental factors such as temperature fluctuations, humidity, and electromagnetic interference can introduce noise and complicate data interpretation. The large volume of data generated by continuous monitoring necessitates robust processing, storage, and analysis frameworks. Additionally, integrating SHM into existing structures can be complex, requiring retrofitting strategies and minimal disruption. Addressing these challenges requires multidisciplinary approaches, combining material science, sensor technology, signal processing, and data analytics.

The future of SHM lies in intelligent, autonomous systems that integrate smart materials, wireless sensor networks, and machine learning. Advances in metamaterials and wave-guiding structures will enable subwavelength detection, higher sensitivity, and selective monitoring of critical components. Predictive models based on historical data and real-time measurements will facilitate condition-based maintenance and failure prediction. Miniaturized, energy-efficient, and self-powered sensors will expand the applicability of SHM in remote or inaccessible locations. The combination of adaptive sensing, real-time analytics, and advanced material engineering promises to revolutionize the monitoring, maintenance, and safety of complex structures and machinery.

4.5.1 Fundamentals of Structural Health Monitoring

At its core, SHM involves measuring key parameters that reflect the mechanical behavior of a structure, such as strain, displacement, vibration, or acoustic emissions. Deviations from expected patterns often indicate the presence of damage, material degradation, or structural defects.

Traditional SHM methods relied on visual inspection or periodic testing, which were limited by accessibility, human error, and the inability to detect internal defects. Advances in sensor technologies, including fiber optics, piezoelectric transducers, and accelerometers, allow for continuous monitoring and high-resolution detection of stress, strain, and vibrational responses. These measurements provide insight into the structural condition and enable predictive maintenance strategies.

4.5.2 Wave-Based Monitoring Techniques

Wave propagation methods have emerged as a powerful tool in SHM due to their ability to detect internal damage over large areas. Ultrasonic waves, guided Lamb waves, and elastic waves are particularly sensitive to cracks, delaminations, corrosion, and other defects. When waves encounter anomalies, they undergo reflection, scattering, or mode conversion, providing identifiable signatures of damage. By analyzing changes in wave amplitude, velocity, phase, or frequency content, engineers can locate and characterize defects. Wave-based SHM techniques offer high sensitivity, rapid assessment, and the ability to monitor complex geometries, making them essential in aerospace, bridge monitoring, and large-scale infrastructure applications.

4.5.3 Role of Sensors and Transducers

Sensors and transducers are the backbone of SHM systems, converting physical quantities into electrical signals for analysis. Piezoelectric sensors, including lead zirconate titanate (PZT) elements, are widely used due to their dual capabilities as actuators and receivers of mechanical waves. Fiber optic sensors, particularly Fiber Bragg Gratings (FBGs), provide high-resolution strain measurements and immunity to electromagnetic interference. Accelerometers, magnetostrictive sensors, and MEMS devices extend the sensing capabilities to vibrations and dynamic responses. The strategic placement and density of these sensors determine the coverage, sensitivity, and reliability of SHM systems. Sensor networks can be embedded into structures during fabrication or retrofitted onto existing structures, enabling continuous monitoring without disrupting operation.

4.5.4 Data Acquisition and Signal Processing

Collecting and interpreting data is a critical aspect of SHM. High-fidelity data acquisition systems capture waveforms, vibration signals, and strain patterns from multiple sensors simultaneously. Signal processing techniques, such as Fast Fourier Transform (FFT), Wavelet Transform, and modal analysis, extract relevant features from raw data, identifying anomalies indicative of structural defects. Advanced methods like time-frequency analysis, correlation analysis, and statistical pattern recognition enhance the detection sensitivity. Machine learning and artificial intelligence are increasingly applied to process large datasets, classify damage types, and predict structural performance, enabling automated, intelligent SHM systems.

4.5.5 Vibration-Based Monitoring

Vibration-based SHM relies on measuring changes in the dynamic properties of a structure, such as natural frequencies, mode shapes, and damping ratios. Structural damage alters stiffness and mass distribution, producing measurable shifts in vibration characteristics. By comparing baseline measurements to current data, damage can be detected and localized. Vibration-based monitoring is particularly effective for bridges, buildings, and mechanical systems, where dynamic loading is significant. Techniques such as modal analysis, frequency response function assessment, and operational modal analysis provide detailed insights into structural integrity, even under operational conditions.

4.5.6 Acoustic Emission Monitoring

Acoustic emission (AE) monitoring detects transient elastic waves generated by the rapid release of energy due to crack growth, fiber breakage, or other structural changes. AE techniques are highly sensitive and can detect damage in its earliest stages, often before visual evidence appears. By analyzing the arrival time, amplitude, and frequency content of acoustic events, engineers can locate damage and assess its severity. AE monitoring is widely used in composite materials, pressure vessels, pipelines, and aerospace components, providing real-time alerts and contributing to proactive maintenance strategies.

4.5.7 Integration with Smart Materials and Metamaterials

The integration of smart materials and metamaterials has enhanced SHM capabilities. Piezoelectric patches, magnetostrictive elements, and shape-memory alloys can serve as both sensors and actuators, enabling active monitoring systems. Mechanical metamaterials and phononic crystals can guide and localize elastic waves with high precision, enhancing damage detection sensitivity and selectivity. By embedding these materials into structural components, SHM systems can exploit wave manipulation, energy localization, and resonance effects to detect defects at subwavelength scales. This synergy between material design and monitoring technology represents a major advance in SHM efficiency and reliability.

4.5.8 Applications in Civil Infrastructure

SHM plays a vital role in civil infrastructure, including bridges, tunnels, dams, and high-rise buildings. Continuous monitoring of load-bearing elements allows engineers to detect cracks, corrosion, joint failures, and foundation settlement before they escalate into major hazards. Wireless sensor networks, embedded fiber optics, and surface-mounted accelerometers enable real-time assessment without interrupting traffic or operation. Data-driven SHM supports predictive maintenance, extending the lifespan of infrastructure, reducing repair costs, and enhancing public safety. Advanced wave-based techniques are particularly valuable for monitoring critical components in bridges and high-stress regions prone to fatigue.

4.5.9 Applications in Aerospace and Mechanical Systems

In aerospace engineering, SHM ensures the structural integrity of aircraft, spacecraft, and rotorcraft components subjected to dynamic loading, fatigue, and environmental stress. Embedded sensors and guided wave monitoring detect delaminations, microcracks, and material degradation in composite wings, fuselage panels, and turbine blades. Similarly, mechanical systems such as engines, gearboxes, and rotating machinery benefit from vibration and acoustic monitoring, allowing early detection of wear, imbalance, or structural defects. Real-time SHM enhances operational safety, reduces downtime, and supports condition-based maintenance strategies in demanding environments.

CHAPTER 5

Optical and Photonic Metamaterials

5.1 LIGHT CONTROL AT THE NANOSCALE

Light control at the nanoscale is a transformative field that merges physics, materials science, and engineering to unlock unprecedented optical capabilities. Through plasmonics, photonic crystals, metamaterials, and near-field optics, scientists have achieved confinement, guidance, and manipulation of light beyond classical limitations. These advances fuel applications in imaging, communication, sensing, and energy harvesting, redefining what is possible in both fundamental research and practical technology. Although challenges remain in fabrication, losses, and scalability, the future promises continued breakthroughs driven by emerging materials, hybrid systems, and quantum technologies. Ultimately, the mastery of nanoscale light control represents a cornerstone for the next generation of optical science and technology.

Despite significant progress, controlling light at the nanoscale presents challenges. Fabrication of precise nanostructures requires advanced techniques such as electron-beam lithography, focused ion beam milling, and self-assembly, which can be costly and time-consuming. Material limitations, such as high optical losses in metals used for plasmonics, restrict efficiency and performance. Scaling up nanoscale devices for commercial use remains difficult due to fabrication complexity and integration challenges. Additionally, environmental factors such as temperature, contamination, and material degradation can alter nanoscale optical responses, reducing stability and reliability in practical applications.

Future advancements in nanoscale light control will rely on new materials, fabrication methods, and hybrid systems. Two-dimensional materials such as graphene and transition metal dichalcogenides offer unique optical properties, including strong confinement, tunability, and nonlinear responses. Topological photonic systems promise robust light transport immune to defects, enhancing reliability. Hybrid integration of plasmonics, photonics, and metamaterials will create multifunctional devices capable of simultaneous sensing, communication, and energy conversion. Advances in quantum nanophotonics will enable scalable quantum communication and computing platforms. As fabrication technologies mature and computational design methods improve, light control at the nanoscale will continue to push the boundaries of science and engineering.

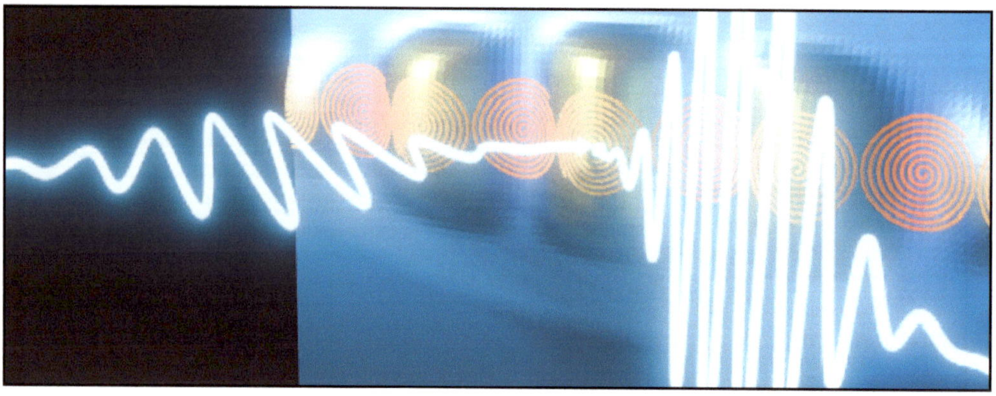

Advancements in Nanoscale Light Control

5.1.1 Fundamentals of Nanoscale Light Interactions

Light–matter interaction at the nanoscale differs significantly from macroscopic optics due to the comparable size of structural features and optical wavelengths. Classical diffraction limits the confinement of light in conventional optics, restricting resolution and wave manipulation. At the nanoscale, however, phenomena such as localized surface plasmon resonances, near-field coupling, and photonic band gaps allow confinement and guidance of light well below the diffraction limit. The use of engineered nanostructures, including nanoparticles, nanowires, and metamaterials, enables tailoring of light propagation, scattering, and absorption with unprecedented precision. These effects are essential for enhancing optical fields, concentrating light energy, and directing photon flows in nanoscale devices.

5.1.2 Plasmonics and Surface Plasmons

Plasmonics plays a central role in controlling light at the nanoscale by exploiting surface plasmon polaritons, which are collective oscillations of electrons at metal–dielectric interfaces. These plasmons couple with incident light, creating confined electromagnetic fields at subwavelength dimensions. Surface plasmons enhance local field intensities by several orders of magnitude, enabling applications in sensing, nonlinear optics, and nanoscale imaging.

Plasmonic nanostructures such as nanoantennas, metallic gratings, and nanoparticle arrays act as efficient light concentrators, channeling optical energy into nanoscale hotspots. This capability is particularly valuable in surface-enhanced Raman spectroscopy, single-molecule detection, and energy harvesting. By tuning the geometry, material composition, and arrangement of plasmonic structures, the resonance frequency and confinement properties can be precisely controlled.

5.1.3 Photonic Crystals and Band Gap Engineering

Photonic crystals represent another approach to nanoscale light control, using periodic dielectric structures to manipulate photon propagation in ways analogous to electron control in semiconductors. These structures create photonic band gaps, frequency ranges where light cannot propagate.

By introducing defects or waveguides within the periodic lattice, it is possible to localize and guide light with extreme precision. Photonic crystals provide opportunities for constructing ultra-compact optical circuits, efficient light emitters, and nanoscale filters. Their ability to suppress or redirect spontaneous emission enhances the performance of lasers, LEDs, and quantum light sources. Band gap engineering allows tailoring of transmission and reflection properties, enabling precise control over optical pathways at the nanoscale.

5.1.4 Metamaterials for Nanoscale Optics

Metamaterials extend nanoscale light control by providing artificial structures with tailored electromagnetic responses not found in natural materials. By engineering subwavelength resonators and unit cells, metamaterials can achieve negative refractive indices, hyperbolic dispersion, and cloaking effects.

These properties enable manipulation of light in unprecedented ways, such as bending light backward, focusing below the diffraction limit, or rendering objects invisible. Metamaterials also facilitate superlensing and near-field imaging, overcoming the limitations of conventional optics. At the nanoscale, their versatility makes them ideal for designing compact optical devices, waveguides, and filters that operate across visible, infrared, and terahertz frequencies.

5.1.5 Near-Field Optics and Localization of Light

Near-field optics is central to nanoscale light control, as it relies on evanescent waves that decay exponentially with distance from a surface but carry subwavelength information. Near-field scanning optical microscopy (NSOM) exploits this principle to achieve resolution beyond the diffraction limit, enabling imaging of structures at the nanometer scale. Localized near-field interactions allow trapping and manipulation of light within nanoscale volumes, enhancing optical forces and enabling applications in nanomanipulation, optical tweezers, and quantum communication. The ability to confine light into regions much smaller than its wavelength provides extraordinary opportunities for enhancing energy densities, improving nonlinear optical effects, and controlling quantum states of light.

5.1.6 Applications in Imaging and Microscopy

Nanoscale control of light has revolutionized imaging and microscopy, enabling scientists to observe phenomena previously inaccessible. Techniques such as super-resolution microscopy, plasmon-enhanced imaging, and near-field optical microscopy allow visualization of molecular and subcellular structures beyond classical diffraction limits. These advances are crucial in biology, medicine, and nanotechnology, where resolving features at the nanoscale provides insights into disease mechanisms, material behavior, and nanoscale assembly processes. Furthermore, the integration of metamaterials and plasmonics into imaging systems enhances resolution, sensitivity, and contrast, paving the way for breakthroughs in diagnostics and molecular imaging.

5.1.7 Applications in Communication and Computing

The control of light at the nanoscale is essential for next-generation communication and computing technologies. Photonic circuits, which use photons instead of electrons, require precise nanoscale waveguides, modulators, and couplers to process information. Plasmonic and photonic crystal components provide pathways for achieving high-speed, low-loss optical data transmission at nanometer scales. Integration of nanoscale optics with silicon photonics supports the development of compact, energy-efficient communication systems. Additionally, light control at the nanoscale is central to quantum computing and secure communication, where quantum dots, single-photon emitters, and nanoscale resonators enable manipulation of quantum information carriers.

5.1.8 Applications in Sensing and Detection

Nanoscale light control is particularly powerful in sensing, where enhanced light–matter interactions improve sensitivity and selectivity. Plasmonic sensors exploit localized field enhancements to detect minute changes in refractive index, enabling label-free detection of biomolecules, pathogens, and pollutants. Photonic crystal sensors achieve wavelength-specific detection through shifts in band structure, while metamaterial-based sensors provide tunable responses to a wide range of physical and chemical stimuli. These nanoscale sensing technologies have applications in medical diagnostics, environmental monitoring, and industrial quality control, offering real-time, ultrasensitive detection capabilities.

5.1.9 Applications in Energy Harvesting

Harnessing light at the nanoscale is crucial for improving energy harvesting technologies, including solar cells and photocatalysis. Plasmonic nanoparticles enhance light absorption by concentrating electromagnetic fields in active regions of solar devices, increasing their efficiency. Photonic crystals and metamaterials provide light-trapping mechanisms that maximize absorption across broad spectral ranges. Additionally, nanoscale engineering enables selective control over photon absorption and emission, optimizing energy transfer processes. These advancements contribute to the development of next-generation photovoltaic devices, photothermal systems, and photocatalytic reactors that leverage nanoscale optics for sustainable energy solutions.

5.2 PLASMONIC AND DIELECTRIC METAMATERIALS

Plasmonic and dielectric metamaterials represent two fundamental strategies for controlling light at subwavelength scales, each with distinct strengths and limitations. Plasmonics harness electron oscillations in metals to achieve extreme field confinement and enhancement, ideal for sensing and imaging, but limited by high losses. Dielectric metamaterials, by contrast, offer low-loss, highly efficient manipulation of wavefronts, enabling flat optical devices and compact photonic components. Hybrid approaches that combine the two promise multifunctional systems that balance sensitivity and efficiency. Together, these metamaterials form the backbone of nanoscale optics, paving the way for innovations in communication, energy, imaging, and sensing that will shape the future of technology.

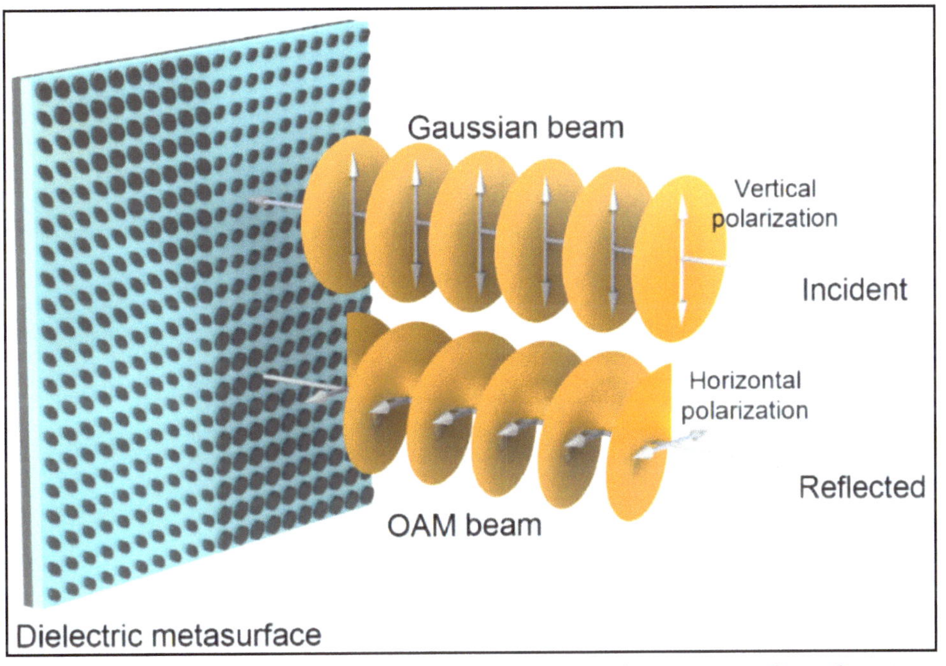

Plasmonic and Dielectric Metasurfaces—Design, Fabrication and Applications

The future of plasmonic and dielectric metamaterials lies in their convergence and integration with emerging technologies. In quantum optics, these materials provide platforms for manipulating single photons and enhancing light-matter coupling. In communication, they support ultra-compact and efficient photonic circuits. In medicine, they enable high-resolution imaging, targeted therapy, and advanced diagnostics. As computational design tools such as machine learning and inverse design accelerate the discovery of novel nanostructures, the performance and versatility of metamaterials will continue to improve. The synergy of plasmonic and dielectric approaches is likely to drive breakthroughs in nanoscale optics, leading to transformative devices across science, industry, and healthcare.

5.2.1 Plasmonic Metamaterials

Plasmonic metamaterials are based on the principle of surface plasmon resonances, which are collective oscillations of free electrons at the interface between metals and dielectrics. When light interacts with these structures, the electromagnetic field couples to electron oscillations, producing confined modes that are localized at the surface. This confinement enables light manipulation below the diffraction limit, a critical advantage for nanoscale optics.

Plasmonic materials typically employ metals such as gold, silver, or aluminum, which support strong plasmonic responses in the visible and near-infrared regions. The strength and frequency of the plasmonic resonance depend on the material, geometry, and arrangement of the nanostructures, allowing tunability across a wide spectral range.

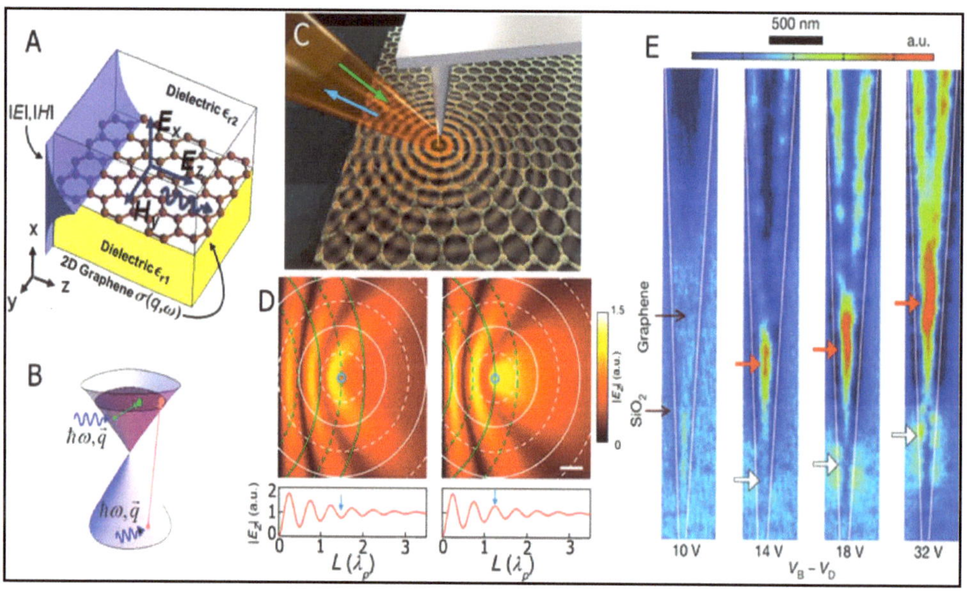

Plasmonic Metamaterials

Advantages and Applications

Plasmonic metamaterials excel at concentrating light into extremely small volumes, producing intense local fields. This property is particularly useful in surface-enhanced spectroscopies such as Raman and fluorescence, where detection of single molecules becomes feasible. Plasmonic waveguides enable light transmission through subwavelength channels, facilitating compact optical circuits. Additionally, plasmonic nanostructures serve as antennas that convert freely propagating waves into localized energy, enhancing light-matter interactions. These characteristics have important applications in biosensing, nanoscale imaging, photothermal therapy, and solar energy harvesting. For example, plasmonic nanoparticles embedded in photovoltaic devices increase absorption efficiency by scattering and trapping light within active layers.

Limitations

Despite their advantages, plasmonic metamaterials face limitations due to intrinsic losses. Metals absorb light through electron scattering, leading to significant ohmic losses that limit efficiency and propagation lengths. This drawback poses challenges for practical implementations, particularly in optical communication and energy devices where low-loss transmission is essential. Additionally, the fabrication of precise plasmonic nanostructures requires advanced techniques such as electron-beam lithography, which can be costly and complex for large-scale production. These issues motivate the exploration of alternative strategies, such as hybrid designs or all-dielectric metamaterials, to overcome loss-related challenges.

5.2.2 Dielectric Metamaterials

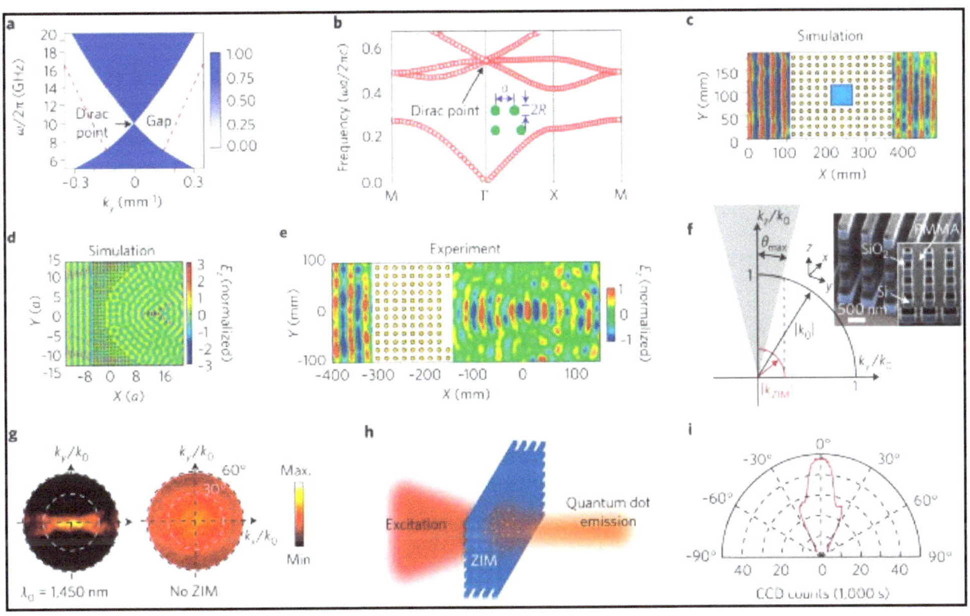

Dielectric Metamaterials

Dielectric metamaterials offer a contrasting approach by relying on high-refractive-index dielectric materials to manipulate electromagnetic waves. Instead of electron oscillations, dielectric metamaterials use Mie resonances and interference effects to control scattering, transmission, and localization of light. Materials such as silicon, gallium nitride, and titanium dioxide are commonly used due to their high refractive indices and low absorption in visible and infrared regions. Unlike plasmonic systems, dielectric metamaterials can achieve strong field confinement with minimal losses, making them highly suitable for practical applications where efficiency is paramount.

Advantages and Applications

Dielectric metamaterials provide low-loss operation across a broad spectral range, enabling high-efficiency optical components. They are particularly effective in manipulating phase, amplitude, and polarization of light, leading to the development of dielectric metasurfaces that function as ultra-thin lenses, holographic projectors, and beam shapers. By precisely engineering the size and arrangement of dielectric nanostructures, it is possible to control wavefronts with subwavelength resolution, replacing bulky optical components with flat, compact devices. Applications extend to augmented reality displays, imaging systems, and light detection and ranging (LiDAR). Dielectric metamaterials are also important in nonlinear optics, where their low losses and strong field enhancement enable efficient frequency conversion and optical switching.

5.2.3 Comparison between Plasmonic and Dielectric Metamaterials

Plasmonic and dielectric metamaterials represent complementary approaches to light control at the nanoscale. Plasmonic systems excel at extreme field localization and enhancement, making them ideal for sensing and applications where strong light–matter interaction is needed. However, they suffer from high losses that limit long-range propagation. Dielectric metamaterials, on the other hand, provide low-loss operation and efficient wavefront manipulation but do not achieve the same degree of field confinement as plasmonics. In practice, the choice between the two depends on the application: plasmonics for extreme sensitivity and nanoscale field effects, and dielectrics for efficient, large-scale, and practical optical components.

5.2.4 Hybrid Plasmonic–Dielectric Designs

Recent advances have focused on combining plasmonic and dielectric elements to harness the strengths of both systems while mitigating their limitations. Hybrid designs incorporate plasmonic nanostructures with dielectric components to achieve both strong field localization and reduced losses. For instance, dielectric nanoparticles can guide light efficiently, while embedded plasmonic elements enhance local interactions. Such hybrid metamaterials open possibilities for multifunctional devices that integrate efficient transmission, sensing, and nonlinear effects within a single platform. These designs are especially relevant in integrated photonics, quantum optics, and multifunctional nanodevices.

5.2.5 Role in Imaging and Sensing

Both plasmonic and dielectric metamaterials have significantly advanced imaging and sensing technologies. Plasmonic sensors, based on shifts in resonance frequency caused by changes in the surrounding refractive index, achieve label-free detection of biomolecules at extremely low concentrations. Dielectric metasurfaces, with their ability to manipulate phase and polarization, enable flat optical elements for compact imaging systems, including lenses that surpass diffraction limits. Together, these technologies support high-resolution medical imaging, environmental monitoring, and portable diagnostic tools. By integrating plasmonic sensitivity with dielectric efficiency, next-generation sensors can achieve higher accuracy and broader applicability.

5.2.6 Role in Energy and Photovoltaics

Energy harvesting is another critical application area for plasmonic and dielectric metamaterials. Plasmonic structures enhance light absorption in thin-film solar cells by scattering and concentrating light, increasing photocurrent generation. Dielectric metamaterials, meanwhile, improve light trapping and reduce reflection losses in photovoltaic devices. The combination of plasmonic nanoparticles for local field enhancement and dielectric structures for broadband light management enables highly efficient solar devices. These approaches also extend to photothermal applications, where concentrated light energy is converted into heat for water purification, cancer therapy, and thermophotovoltaic systems.

5.3 METASURFACES AND FLAT OPTICS

Metasurfaces and flat optics represent a groundbreaking shift in the way we understand and design optical systems. By harnessing engineered nanostructures to control light in ultrathin layers, metasurfaces overcome the limitations of traditional bulky optics. They enable compact, lightweight, and multifunctional devices with extraordinary capabilities, ranging from high-resolution imaging to holography and communication. While challenges in efficiency, scalability, and integration remain, ongoing advancements are rapidly addressing these issues. As research progresses, metasurfaces are set to become a cornerstone of future optical technologies, transforming the way light is controlled and applied in everyday life.

Metasurfaces can be understood as the two-dimensional analog of metamaterials. While metamaterials are artificially structured composites that manipulate electromagnetic waves through three-dimensional volumetric design, metasurfaces compress this concept into a single layer of nanostructures. These nanoscale building blocks, often referred to as meta-atoms, are engineered to impose specific effects on incoming waves. By altering their size, geometry, arrangement, and orientation, each element introduces a local phase shift or polarization modification, enabling precise control of the optical response. The collective action of these engineered scatterers produces tailored wavefronts, granting metasurfaces the ability to perform functions such as focusing, steering, and holography in a way that is both efficient and compact.

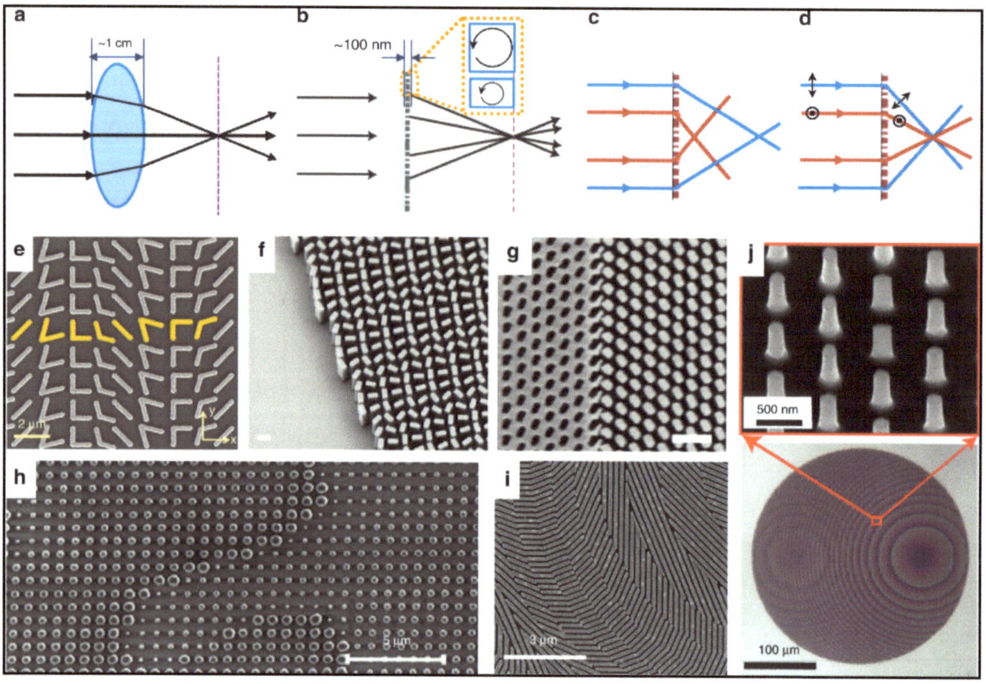

Metasurfaces

Principles of Operation

The key principle behind metasurfaces lies in their capacity to manipulate the fundamental parameters of light. Light waves can be described in terms of amplitude, phase, frequency, and polarization, and metasurfaces provide unprecedented control over each of these. For example, by designing nanostructures to create abrupt phase discontinuities, a metasurface can reshape an incoming planar wave into a focused beam, mimicking the behavior of a lens without requiring curvature or thickness. Similarly, metasurfaces can convert linear polarization into circular polarization or generate beams with orbital angular momentum. Unlike conventional optics that rely on the gradual accumulation of phase during propagation, metasurfaces achieve complete phase control within a subwavelength thickness, which makes them exceptionally powerful tools for flat optics.

Flat Optics and Its Significance

Flat optics is the discipline that develops planar optical components using metasurfaces instead of bulky traditional devices. Its importance lies not only in the reduction of size and weight but also in the enhancement of optical performance. By replacing thick multi-element lens systems with a single flat surface, optical devices can become more compact and efficient. This has critical implications for portable electronics, medical imaging, and space technologies where size and weight are significant constraints. Additionally, flat optics opens the door to multifunctional devices that can integrate several optical functions into one layer, something that is not easily achievable with conventional optics.

Metalenses and Imaging Applications

One of the most revolutionary applications of metasurfaces is in the creation of flat lenses, also called metalenses. These lenses use arrays of nanostructures to focus light with exceptional accuracy. Metalenses can be designed to eliminate common problems such as spherical and chromatic aberrations that limit the performance of traditional lenses. Achromatic metalenses, for instance, can bring multiple wavelengths of light to the same focal point, providing sharper and clearer images.

This has profound applications in microscopy, medical endoscopy, and photography, where smaller, lighter, and higher-performing lenses are increasingly in demand. Moreover, the ability to manufacture these lenses using semiconductor fabrication techniques suggests their integration into compact consumer devices such as smartphones and augmented reality systems.

Wavefront Shaping and Holography

Beyond focusing, metasurfaces have the capacity to engineer entire wavefronts with high precision. This capability has been used to create holograms that are thinner, lighter, and more efficient than those produced using conventional methods. Metasurface-based holography relies on encoding the phase and amplitude information into nanostructures, thereby reconstructing three-dimensional images with great fidelity. This advancement has significant implications for augmented and virtual reality, display technologies, and secure data storage. Furthermore, metasurfaces can generate complex beam profiles such as vortex beams, which carry orbital angular momentum and can be used for advanced communication systems and particle manipulation.

Polarization Control and Spin-Orbit Coupling

Another domain where metasurfaces excel is polarization control. By designing anisotropic nanostructures, metasurfaces can manipulate the polarization of light in ways that traditional optics cannot easily achieve. For example, metasurfaces can function as ultrathin waveplates, polarizers, or converters that alter the handedness of circularly polarized light. They also enable the study and utilization of spin-orbit interactions, where the polarization state of light influences its spatial distribution. Such control is essential in quantum optics, secure communication, and advanced imaging techniques, where polarization plays a critical role in information encoding and retrieval.

Nonlinear and Active Metasurfaces

While much of metasurface research has focused on linear interactions, recent advances have extended into nonlinear optics. Nonlinear metasurfaces exploit strong light–matter interactions at the nanoscale to generate higher harmonics, frequency mixing, or optical switching effects. These developments enable ultracompact devices for frequency conversion, optical computing, and high-speed communication. Additionally, active metasurfaces incorporating tunable materials such as graphene, liquid crystals, or phase-change compounds allow dynamic control over their optical properties. These reconfigurable metasurfaces pave the way for adaptive optics, programmable holography, and real-time beam steering, offering versatility that goes far beyond traditional optical devices.

Fabrication Techniques and Challenges

The development of metasurfaces is closely tied to advancements in nanofabrication. Techniques such as electron-beam lithography, nanoimprinting, and advanced photolithography are used to create the intricate nanostructures that define metasurface behavior. However, scalability remains a challenge, as producing large-area metasurfaces with nanoscale precision can be costly and time-consuming. Material losses, especially in the visible spectrum, also limit efficiency, while ensuring robustness and durability under real-world conditions remains an area of active research. Addressing these challenges is crucial for the widespread commercialization of flat optics.

5.4 SUPERLENSING AND NEAR-FIELD IMAGING

Superlensing and near-field imaging represent one of the most exciting advancements in modern optics, providing a way to overcome the long-standing diffraction limit. By recovering evanescent waves and reconstructing subwavelength details, superlenses unlock new possibilities for imaging, fabrication, and communication at the nanoscale. From semiconductor technology to biomedical diagnostics, the impact of these innovations is already evident and will continue to grow as new materials and fabrication techniques mature. While challenges in efficiency, scalability, and wavelength versatility remain, the trajectory of research suggests that superlensing will become a cornerstone of future optical science and technology.

The future of superlensing and near-field imaging is highly promising. Ongoing research aims to design broadband and tunable superlenses that can operate across multiple wavelengths with minimal loss. The integration of superlensing with computational imaging techniques and machine learning could further enhance performance, enabling real-time reconstruction of high-resolution images. Flexible and wearable superlenses are also being explored for applications in biomedical monitoring and portable sensing. Additionally, the combination of superlensing with quantum optics holds the potential to achieve ultra-precise imaging at the single-photon level, revolutionizing microscopy and metrology.

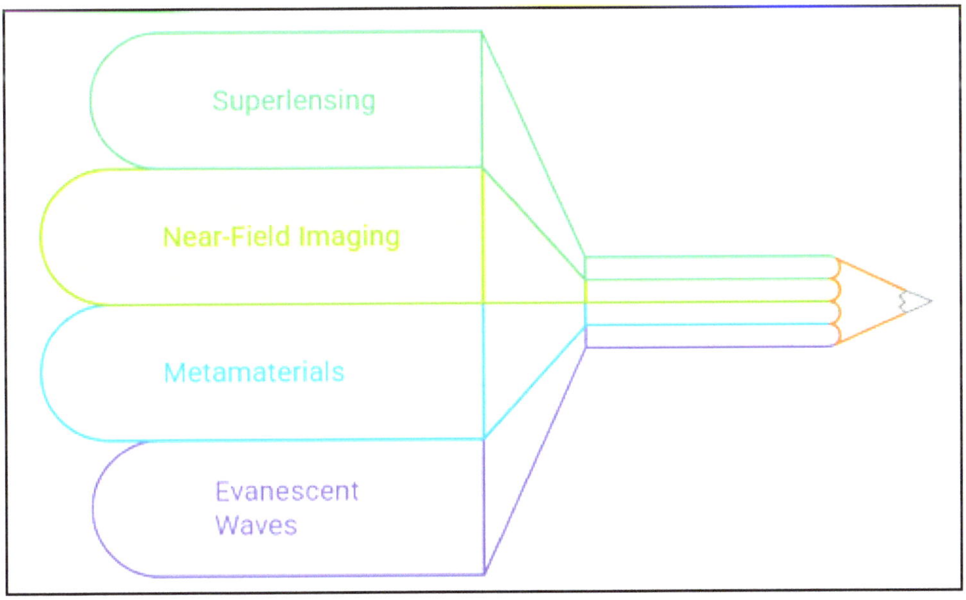

Pushing Optical Boundaries

5.4.1 Diffraction Limit in Conventional Optics

Conventional optical imaging is bound by the principles of wave propagation. When light interacts with an object, it produces both propagating waves and evanescent waves. Propagating waves carry information that travels to the far field and can be captured by lenses, while evanescent waves contain subwavelength details but decay rapidly with distance. The diffraction limit, first formalized by Ernst Abbe in the nineteenth century, states that no optical system can resolve details smaller than roughly half the wavelength of the light used. For visible light, this corresponds to a resolution of around 200 nanometers, which restricts direct observation of many nanoscale structures, such as viruses, proteins, or semiconductor features. Traditional methods such as electron microscopy circumvent this limit but often require destructive sample preparation. Superlensing offers a fundamentally optical solution by directly recovering lost evanescent information.

5.4.2 Superlensing

Superlensing relies on the unique properties of metamaterials to amplify and recover evanescent waves. Unlike conventional lenses that only focus propagating waves, a superlens captures both propagating and non-propagating components of light. This enables it to reconstruct an image with details well below the diffraction limit. The idea was first introduced by Sir John Pendry in 2000, who proposed that a slab of negative refractive index material could act as a perfect lens. Such a lens would not only focus propagating waves but also restore decaying evanescent waves, thereby producing an image with theoretically unlimited resolution. While practical challenges prevent the realization of a truly perfect lens, the principle has been successfully demonstrated in multiple experiments using metamaterials and plasmonic structures.

5.4.3 Negative Refractive Index and Its Role

The key enabler of superlensing is the phenomenon of negative refraction. In materials with a negative refractive index, the direction of light propagation is reversed compared to normal materials. This unusual behavior allows such materials to bend light in ways not possible with natural substances. When used in lensing, negative refraction enables the amplification of evanescent waves. The material essentially recovers subwavelength details that would otherwise vanish, restoring them to the image plane. Early demonstrations of superlenses used silver films at ultraviolet frequencies because silver supports surface plasmons that interact strongly with evanescent fields. Later, engineered metamaterials expanded the concept into other spectral ranges, including microwave, terahertz, and infrared.

5.4.4 Near-Field Imaging Principles

Near-field imaging refers to techniques that exploit the evanescent field information present in the immediate vicinity of an object. Because evanescent waves decay exponentially with distance, detectors or lenses must be placed extremely close to the object to capture them. Superlenses act as bridges, extending the reach of evanescent waves so that they can contribute to image formation. This near-field amplification is what enables imaging beyond the diffraction limit. Unlike far-field techniques that rely on propagating waves, near-field imaging captures the rich detail encoded in localized fields, making it indispensable for exploring nanostructures and surface phenomena.

5.4.5 Superlens Designs and Implementations

The design of superlenses has evolved significantly since Pendry's proposal. The simplest implementation is the planar silver superlens, which uses a thin metallic film to enhance evanescent waves through surface plasmon resonance. This design successfully demonstrated sub-diffraction imaging in the ultraviolet regime. More advanced designs include hyperbolic metamaterial lenses, which use anisotropic structures to support high-momentum modes and extend the range of captured details. Another approach is the use of hyperlenses, which transform evanescent waves into propagating waves by employing cylindrical or spherical geometries. These devices project nanoscale details into the far field, enabling conventional detectors to capture high-resolution images. Each of these designs represents a step toward practical applications of superlensing in different spectral domains.

5.4.6 Applications in Nanotechnology

Superlensing has had a profound impact on nanotechnology, where the ability to resolve and manipulate features at the nanoscale is critical. In semiconductor manufacturing, for example, superlenses offer new methods for lithography and defect inspection. Traditional optical lithography struggles with feature sizes approaching the diffraction limit, but superlenses can provide finer patterning without resorting to expensive alternatives such as extreme ultraviolet light sources. Similarly, superlensing enhances the ability to monitor nanoscale fabrication processes in real time, improving precision and yield in microelectronics production.

5.4.7 Applications in Biology and Medicine

In the life sciences, superlensing has transformed imaging of biological structures. Traditional optical microscopes cannot resolve features smaller than 200 nanometers, which excludes many viruses, protein complexes, and cellular structures. Superlenses enable optical visualization of these nanoscale entities without the need for invasive labeling or destructive electron microscopy. For example, silver superlenses have been used to image subcellular features with remarkable clarity. This opens the door to advances in medical diagnostics, where observing disease markers at the nanoscale can lead to earlier and more accurate detection. The non-invasive nature of superlensing also makes it suitable for studying living cells and dynamic biological processes.

5.4.8 Advances in Communication and Data Storage

Superlensing also plays a role in optical communication and data storage, where increasing demand for miniaturization and bandwidth requires resolving and controlling nanoscale features. In optical data storage systems, superlenses can enable higher density recording by focusing light into spots smaller than the diffraction limit allows. This could significantly expand the capacity of optical discs and other storage media. In communication technologies, superlensing can enhance the efficiency of photonic circuits by enabling precise coupling between nanoscale components, leading to faster and more compact devices for processing information.

5.5 TUNABLE AND ACTIVE OPTICAL METAMATERIALS

Tunable and active optical metamaterials mark a significant evolution from static designs, offering dynamic and reconfigurable control over light. Through mechanisms such as electrical, thermal, optical, and mechanical modulation, these systems enable adaptive lenses, reconfigurable holograms, dynamic filters, and real-time sensing platforms. The integration of advanced materials like graphene, liquid crystals, and phase-change compounds further expands the possibilities for functionality. Although challenges remain in efficiency, scalability, and durability, rapid progress in material science and nanofabrication is addressing these limitations. As research continues, tunable and active metamaterials will reshape optics, leading to transformative applications in imaging, communication, computing, and beyond.

5.5.1 Tunable Metamaterials

Tunable metamaterials are structures whose optical properties can be deliberately modified after fabrication. Unlike static metamaterials, which exhibit a fixed response based on geometry and composition, tunable metamaterials incorporate elements that respond to electrical, optical, thermal, or mechanical stimuli. This adaptability enables them to function across broader frequency ranges, adjust to changing environmental conditions, or switch between different functionalities. For example, a tunable lens could shift its focal length under an applied voltage, or a reconfigurable filter could adjust its transmission spectrum to accommodate different wavelengths. Such versatility makes tunable metamaterials highly attractive for next-generation photonics.

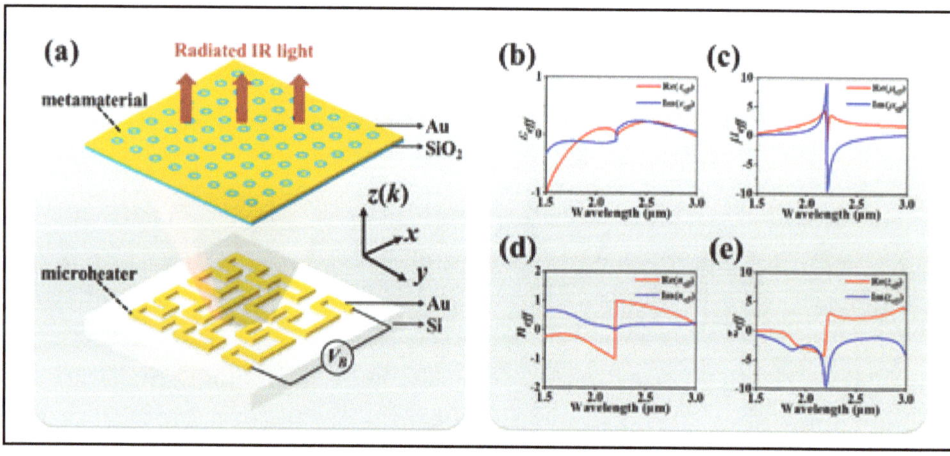

Tunable Metamaterials

5.5.2 Active Metamaterials and their Distinction

Active metamaterials are a subset of tunable designs that not only adjust passively but also integrate gain or energy sources to amplify or compensate for losses. Traditional metamaterials, especially those based on metals, often suffer from energy dissipation due to absorption. By incorporating active elements such as semiconductor quantum wells, laser dyes, or nonlinear media, these structures can inject energy back into the system. This leads to enhanced performance, including higher efficiency, controllable amplification, or even lasing actions. The distinction between tunable and active metamaterials lies in the source of control: tunable systems respond to external modulation, while active systems embed energy sources for continuous or dynamic feedback.

5.5.3 Mechanisms of Tunability

Several mechanisms enable tunability in optical metamaterials. One widely studied method is electrical tuning, where applying a voltage alters the properties of integrated elements such as graphene layers, liquid crystals, or electro-optic materials. This approach provides fast and precise control over phase and amplitude. Thermal tuning is another mechanism, where temperature changes induce modifications in refractive index or phase-change materials, such as vanadium dioxide or germanium-antimony-tellurium alloys.

Optical tuning uses incident pump light to alter the optical state of nonlinear materials, enabling ultrafast switching and modulation. Mechanical tuning involves physically reconfiguring the metamaterial structure, often through micro-electromechanical systems (MEMS) or flexible substrates, to dynamically change the arrangement of the nanostructures. Each mechanism offers distinct advantages depending on the application and desired level of control.

5.5.4 Graphene-Based Tunable Metamaterials

Graphene has emerged as a particularly powerful material for tunable and active optical metamaterials. Its unique electronic band structure allows for adjustable conductivity through electrostatic gating, making it ideal for broadband and ultrafast modulation.

Graphene-integrated metamaterials can dynamically control absorption, reflection, and transmission across terahertz to infrared frequencies. For instance, graphene-based absorbers have been demonstrated to achieve near-perfect tunability, while graphene-loaded metasurfaces enable adaptive beam steering and dynamic holography. The combination of graphene's exceptional carrier mobility, mechanical flexibility, and transparency positions it as a cornerstone for future active optical devices.

5.5.5 Phase-Change Materials in Active Metamaterials

Phase-change materials represent another key strategy for tunable metamaterials. Compounds such as vanadium dioxide and chalcogenide alloys exhibit reversible transitions between insulating and metallic or crystalline and amorphous phases when subjected to thermal, electrical, or optical excitation. These transitions lead to dramatic changes in optical properties, including refractive index and absorption coefficient. Integrating phase-change materials into metasurfaces enables reconfigurable lenses, optical switches, and programmable holograms. A single device can shift between transparent and opaque states, alter its focusing properties, or switch polarization responses, offering multifunctionality in ultrathin formats.

5.5.6 Liquid Crystal Metasurfaces

Liquid crystals, already widely used in display technologies, have found a new role in tunable metamaterials. By embedding metasurfaces within liquid crystal layers, researchers can achieve reconfigurability controlled by applied voltages. Liquid crystals respond by reorienting their molecules, which modifies the effective refractive index surrounding the nanostructures.

This leads to tunable phase delays and polarization effects. Liquid crystal metasurfaces are attractive for applications in adaptive lenses, beam steering, and reconfigurable optical filters. Their relatively low power consumption and mature fabrication processes make them promising candidates for near-term commercialization.

5.5.7 Nonlinear Optical Tunability

Nonlinear effects also provide powerful avenues for tunability in metamaterials. By using materials with strong nonlinear responses, such as semiconductors, dielectrics, or plasmonic structures, metasurfaces can alter their behavior depending on the intensity of incident light.

For instance, nonlinear metasurfaces can exhibit harmonic generation, all-optical switching, and intensity-dependent refractive index changes. These properties are particularly useful for ultrafast communication technologies, where modulation speeds must exceed those achievable with electrical or thermal methods. Nonlinear tunable metamaterials therefore form a bridge between nanophotonics and high-speed optical signal processing.

5.5.8 Applications in Imaging and Sensing

Tunable and active optical metamaterials have opened new possibilities in imaging and sensing technologies. In imaging, adaptive metalenses can dynamically change focal length or correct aberrations, enabling compact zoom lenses or real-time focusing adjustments without mechanical movement.

In sensing, tunable metasurfaces allow for selective detection of chemical and biological species by adjusting resonant frequencies to match specific molecular signatures. Active sensing platforms can even amplify weak signals, improving detection sensitivity for trace elements or low-concentration biomolecules. These capabilities are particularly important in medical diagnostics, environmental monitoring, and security applications.

5.5.9 Applications in Communication and Data Processing

In communication systems, tunable metamaterials are vital for managing the increasing demand for bandwidth and efficiency. Dynamically reconfigurable antennas and metasurfaces can adaptively steer beams, reduce interference, and optimize data transmission across varying conditions. Active metamaterials also enable modulators and switches that operate at high speeds for optical communication networks. In data processing, tunable metasurfaces are being investigated for optical computing platforms, where they can perform logic operations, information routing, and wavelength multiplexing in ultracompact formats. Such integration could lead to faster and more energy-efficient computing systems.

CHAPTER 6

Thermal and Thermo-Mechanical Metamaterials

6.1 HEAT CLOAKING AND THERMAL CAMOUFLAGE

Heat cloaking and thermal camouflage represent groundbreaking advancements in the field of thermal metamaterials. By manipulating heat conduction and infrared emission, these technologies offer new ways to hide, protect, and manage thermal signatures.

From military applications that enhance stealth capabilities to civilian uses in electronics, architecture, and energy management, their potential is vast. While challenges such as scalability, efficiency, and durability remain, rapid progress in material science and design methodologies continues to push the boundaries of what is possible. Ultimately, heat cloaking and thermal camouflage illustrate the transformative power of metamaterials in reshaping the way we interact with fundamental physical phenomena.

Despite significant advances, several challenges remain in the development of heat cloaking and thermal camouflage. One major difficulty is achieving broadband and omnidirectional performance. Many current cloaking systems work only within limited temperature ranges or directions of heat flow. Scalability is another concern, as most demonstrations are limited to laboratory-scale prototypes. Integrating these technologies into large objects such as vehicles or buildings requires advances in fabrication methods and cost-effective materials. Additionally, active systems that rely on external power must be optimized for energy efficiency to be practical in real-world applications.

6.1.1 Concept of Heat Cloaking

Heat cloaking is based on the principle of transformation thermodynamics, which borrows ideas from transformation optics. In transformation optics, electromagnetic waves can be guided around an object, rendering it invisible to detection. Analogously, in transformation thermodynamics, heat conduction paths are manipulated so that thermal energy flows around an object rather than through it. The key to heat cloaking lies in designing metamaterials with anisotropic thermal conductivities that redirect heat flux. When properly engineered, the cloak ensures that the temperature distribution outside remains undisturbed, making the hidden object thermally invisible. This allows an object to be shielded from detection by infrared cameras or sensors that rely on heat gradients.

6.1.2 Mechanisms of Thermal Cloaks

Thermal cloaks can be designed through layered structures, gradient-index materials, or metastructures with carefully engineered thermal conductivity. One approach uses concentric shells of materials with different thermal conductivities, creating a path for heat to flow around the central region. Another method involves gradient-index designs where the effective thermal conductivity varies continuously across the cloak. Metastructures such as phononic crystals and composite lattices can also be used to mimic anisotropic thermal properties. Importantly, these designs require no active power input; instead, they rely on the passive redirection of heat flow. More advanced active cloaks use external controls, such as voltage-driven elements, to dynamically adjust thermal properties, making the cloak adaptable to changing conditions.

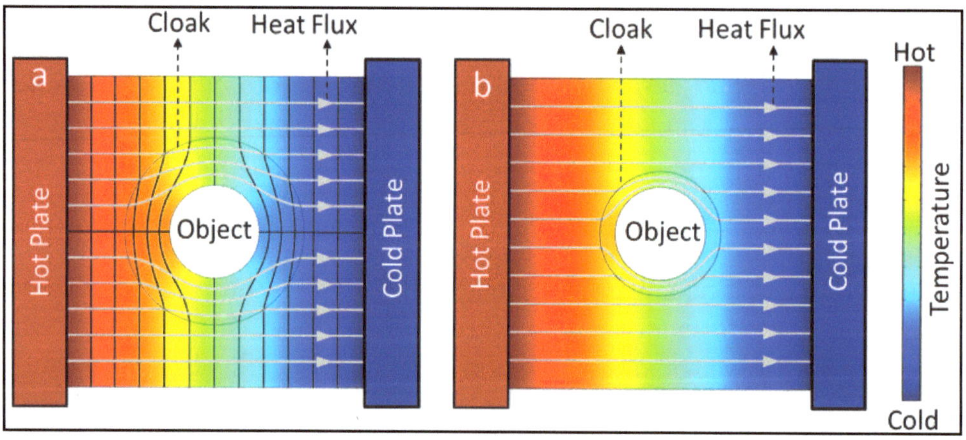

Illustration of Two Approaches of Thermal Invisibility Cloaking

6.1.3 Thermal Camouflage as an Extension

While heat cloaking hides the presence of an object by redirecting heat, thermal camouflage focuses on altering thermal radiation to blend objects into their backgrounds. Every object emits infrared radiation according to its temperature and emissivity. Infrared cameras and thermal sensors exploit this radiation to detect and identify objects. Thermal camouflage aims to manipulate emissivity, either by suppressing the infrared emission or by tuning it to match the environment. Metamaterial-based coatings and metasurfaces can be engineered to provide dynamic control over emissivity. In some cases, thermal camouflage can be combined with cloaking strategies, where both conduction and emission are manipulated to provide a higher degree of invisibility.

6.1.4 Materials for Heat Cloaking and Camouflage

Several materials have been identified as promising candidates for heat cloaking and camouflage. Phase-change materials such as vanadium dioxide are particularly important because they undergo reversible transitions between insulating and metallic states at certain temperatures. This property allows dynamic control over both conductivity and emissivity. Graphene is another key material because of its tunable conductivity and broadband optical absorption, enabling control over infrared radiation. Ceramic composites, dielectric coatings, and multilayer metasurfaces also provide opportunities to design structures with tailored thermal properties. The integration of these materials with metamaterial architectures enhances their ability to provide cloaking or camouflage functionalities across a range of frequencies and conditions.

6.1.5 Experimental Demonstrations

Several experimental breakthroughs have proven the feasibility of heat cloaking and thermal camouflage. Early demonstrations of thermal cloaks used layered rings of copper and polymer to redirect heat around an object. These experiments showed that the external temperature field remained unchanged, effectively hiding the cloaked region. More advanced studies demonstrated dynamic cloaking using active elements that adjust conductivity in real time. In thermal camouflage, metasurface coatings capable of tuning emissivity have been tested successfully, allowing objects to remain undetected under infrared imaging. In some cases, smart camouflage systems have been developed where the surface emissivity automatically adapts to environmental changes.

6.1.6 Applications in Defense and Security

One of the most significant applications of heat cloaking and thermal camouflage is in defense and security. Military vehicles, aircraft, and personnel are often detected using infrared sensors and thermal imaging systems. By integrating cloaking or camouflage technologies, these signatures can be suppressed, making it more difficult for adversaries to detect or track targets. Cloaking systems can also protect sensitive installations from thermal surveillance.

In addition to concealment, thermal camouflage may also be used to create decoys by artificially projecting false thermal signatures, confusing enemy detection systems. Such applications highlight the strategic importance of these technologies in modern warfare.

6.1.7 Applications in Electronics and Thermal Management

Beyond defense, heat cloaking has profound implications in electronics and thermal management. Modern electronic devices, such as microprocessors, suffer from localized hotspots that degrade performance and reliability. Thermal cloaks can guide heat away from sensitive regions without disrupting overall cooling systems, improving device stability and extending lifespan. Similarly, thermal camouflage can be used to regulate infrared emission from devices, enhancing energy efficiency in systems where radiative heat loss must be controlled. For instance, satellites and spacecraft require precise thermal regulation to maintain optimal operating conditions, making adaptive thermal cloaks highly valuable.

6.1.8 Applications in Architecture and Energy Efficiency

In the context of buildings and energy efficiency, thermal camouflage technologies can be applied to control heat emission and improve insulation. Cloaking systems can prevent thermal leakage from structures, making them appear cooler in summer or warmer in winter depending on design. Adaptive emissivity coatings could help regulate indoor temperatures by either reflecting heat away or retaining it, reducing the reliance on active heating and cooling systems. This not only enhances comfort but also reduces energy consumption and contributes to sustainability goals.

6.2 ANISOTROPIC HEAT FLOW AND CONDUCTION CONTROL

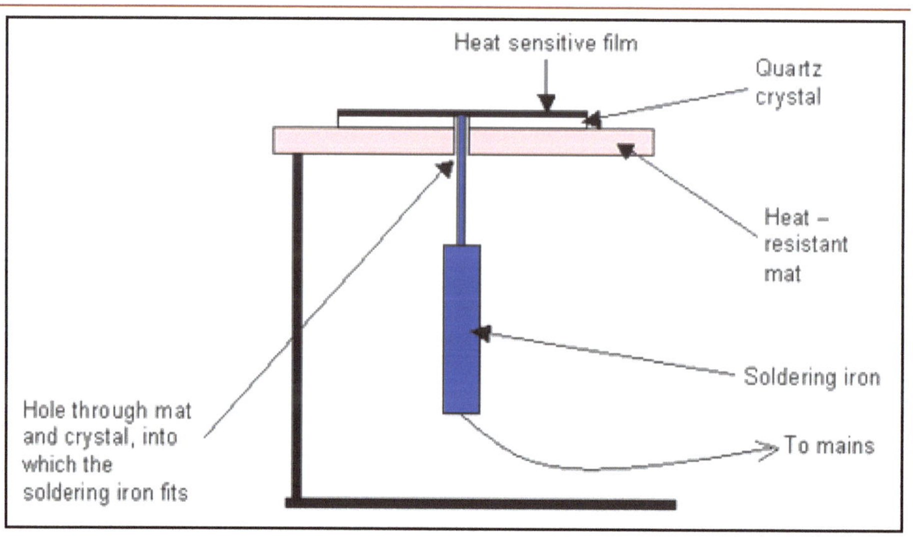

Anisotropic Thermal Conductivity

Heat conduction has always been one of the most fundamental processes in physics and engineering. In conventional materials, the way heat flows is largely dictated by their intrinsic thermal conductivity, which is generally isotropic, meaning the same in all directions. However, in many natural and engineered systems, anisotropy plays a critical role. Anisotropic heat flow refers to a situation where thermal conductivity differs along different directions within a material, enabling preferential guidance of heat. The ability to design and control anisotropic thermal conduction is a cornerstone of modern thermal metamaterials, as it allows for highly specialized applications such as cloaking, energy harvesting, thermal camouflage, and improved heat management in electronics.

Anisotropy in Heat Flow

Anisotropy in thermal conduction arises when a material conducts heat more efficiently along one direction than another. This behavior contrasts with isotropic materials like most metals, where thermal conductivity is uniform in every direction. Natural anisotropy occurs in certain crystals, composites, and layered materials such as graphite, where heat conduction is much higher along the in-plane direction than in the out-of-plane direction. By engineering anisotropy, scientists can tailor the flow of thermal energy in unprecedented ways. This control makes it possible to direct heat around obstacles, concentrate it in specific regions, or suppress it in others, providing a level of functionality that isotropic materials cannot achieve.

Principles of Heat Conduction

Heat conduction is governed by Fourier's law, which states that the heat flux is proportional to the negative gradient of temperature multiplied by the thermal conductivity of the material. In isotropic systems, the thermal conductivity is a scalar quantity, meaning it has a single value. In anisotropic systems, however, thermal conductivity becomes a tensor, with different values along different axes. This tensorial nature enables complex behaviors, such as heat bending around objects or concentrating at designated points. By manipulating the tensor components through engineered structures, it becomes possible to design thermal pathways that achieve highly specific outcomes.

Natural Sources

Anisotropic heat conduction exists in several natural materials. Crystals such as quartz and sapphire exhibit directional differences in thermal conductivity due to their atomic arrangements. Graphite and other layered materials, including transition metal dichalcogenides, display strong anisotropy because their atomic bonds are much stronger in the plane than between the planes. Biological materials, such as bone and wood, also show anisotropic heat conduction because of their hierarchical structures. While these natural examples illustrate the phenomenon, engineered metamaterials allow researchers to greatly amplify and control anisotropy beyond natural limits, expanding its applications in technology.

Engineering Anisotropy through Metamaterials

Metamaterials are artificially designed structures that derive their properties from geometry and arrangement rather than intrinsic composition. Thermal metamaterials can be fabricated with periodic or gradient architectures that create anisotropic heat conduction.

For example, composites made of alternating layers of high- and low-conductivity materials can guide heat preferentially in one direction. More sophisticated designs use complex lattice geometries or microstructured arrays to manipulate conduction tensors. By adjusting the spacing, orientation, and size of these structures, researchers can create effective materials with custom anisotropic properties not found in nature.

Applications in Heat Cloaking

Anisotropic conduction is at the heart of thermal cloaking technologies. By designing materials where heat preferentially flows around a region instead of through it, it is possible to render that region thermally invisible. The cloak's anisotropy ensures that the external temperature field appears undisturbed, while the interior remains shielded from detection. For example, concentric layers of anisotropic composites can bend heat around an object, achieving cloaking effects similar to those in transformation optics. This application demonstrates how anisotropic conduction can be harnessed to manipulate thermal fields for stealth and protection.

Applications in Thermal Lenses

Another exciting application of anisotropic heat flow is in thermal lensing, where heat can be focused or spread out like light through an optical lens. Thermal lenses are designed by engineering anisotropic conductivities such that heat flux converges to a point or diverges from it. This capability allows localized heating for processes like targeted drug delivery or energy harvesting. Similarly, in electronics, thermal lenses can spread heat away from hotspots, preventing localized overheating while maintaining overall performance. Such designs rely on precisely tailored anisotropic tensors to shape thermal fields dynamically.

Role in Energy Harvesting and Conversion

Anisotropic conduction also plays an important role in energy harvesting systems, particularly thermoelectrics. In thermoelectric devices, temperature gradients are converted into electrical energy. Anisotropic materials can optimize the direction of heat flow relative to the direction of electrical conduction, thereby enhancing efficiency. By guiding heat along preferred pathways while suppressing it in others, anisotropic materials allow for improved thermal-to-electrical conversion. In solar thermophotovoltaics, anisotropy can be used to manage heat flow in receivers, concentrating thermal energy in designated zones for improved energy capture.

Applications in Electronics and Thermal Management

Modern electronic devices generate significant heat, particularly in high-performance processors and compact systems. Localized hotspots are a major challenge because they reduce efficiency and reliability. Anisotropic thermal materials offer solutions by guiding heat away from critical components.

For example, composites with high in-plane conductivity but low cross-plane conductivity can spread heat laterally without affecting vertical structures. This approach has been applied to cooling of microchips, flexible electronics, and power devices, where managing heat flow in specific directions is crucial. The ability to tailor anisotropy ensures precise control of temperature distribution across complex systems.

Advanced Materials for Anisotropic Conduction

Several advanced materials are being developed to achieve controlled anisotropy. Graphene and other two-dimensional materials naturally exhibit high in-plane thermal conductivity and can be integrated into composites to enhance directional heat flow. Carbon nanotube networks also provide exceptional anisotropy due to their structural alignment. Phase-change materials offer dynamic anisotropy by switching between crystalline and amorphous states with different conduction properties. Metasurfaces and 3D-printed lattices represent the frontier of engineered anisotropy, where precise design allows for nearly arbitrary manipulation of conduction pathways.

6.3 PROGRAMMABLE THERMAL BEHAVIOR

The field of thermal management has traditionally relied on passive strategies, where materials conduct, store, or radiate heat according to their fixed properties. While effective for many conventional applications, these approaches lack flexibility and adaptability. The advent of programmable thermal behavior marks a transformative shift, enabling materials and systems to actively control, redirect, and manipulate heat flow on demand.

Programmable thermal systems integrate advanced materials, microstructures, and control mechanisms to achieve dynamic thermal responses. Unlike static thermal materials, programmable systems can adapt to changing environmental conditions, modify temperature distributions, and even perform logical operations using heat as a medium. This capability has significant implications for electronics, energy systems, aerospace, medical devices, and defense technologies.

6.3.1 Programmable Thermal Behavior

Programmable thermal behavior refers to the ability to actively control the propagation of heat in a system through design, external stimuli, or feedback mechanisms. At its core, this concept combines materials science with control engineering, allowing thermal properties such as conductivity, emissivity, and heat capacity to be modulated in real time. For instance, a programmable thermal material might redirect heat away from sensitive components during high-load operation and then restore normal conduction once the system cools. This dynamic control contrasts with traditional thermal materials that respond passively to temperature gradients. Programmable thermal systems can be realized using phase-change materials, active metamaterials, composite architectures, or integrated microelectromechanical systems.

6.3.2 Mechanisms for Programming Thermal Responses

Several mechanisms enable programmable thermal behavior. One approach involves phase-change materials, which undergo reversible transitions between solid, liquid, or crystalline states, each with distinct thermal conductivities. By externally controlling these transitions using temperature, electrical input, or optical stimuli, it is possible to tune heat flow dynamically. Another method relies on anisotropic metamaterials, where the thermal conductivity tensor can be adjusted by mechanically reconfiguring microstructures, such as stretchable lattices or MEMS-based systems. Electrical and optical controls are also widely used; for example, applying voltage to graphene or vanadium dioxide layers can modulate thermal conductivity and emissivity. By combining these mechanisms, engineers can create materials whose thermal behavior is programmable across time, space, and frequency.

6.3.3 Thermal Metamaterials and Programmable Functionality

Thermal metamaterials have been central to the development of programmable thermal behavior. Metamaterials are engineered with micro- or nanoscale structures that allow properties to be tailored beyond what is achievable in natural materials. By designing the geometry and composition of metamaterials, it is possible to create directional heat flow, concentrate heat in specific regions, or suppress conduction entirely. Programmable thermal metamaterials extend these capabilities by integrating active elements that can be dynamically adjusted. For instance, a metasurface composed of vanadium dioxide patches can switch between insulating and conducting states when electrically stimulated, enabling the redirection of heat flux on demand. Such systems open the possibility for adaptive thermal cloaks, smart cooling surfaces, and energy-efficient building materials.

6.3.4 Applications in Electronics and Microdevices

One of the most compelling applications of programmable thermal behavior is in electronics and microdevices. Modern processors, power devices, and integrated circuits generate substantial localized heat, which can degrade performance and reduce lifespan. Traditional cooling methods, such as heatsinks or passive thermal interface materials, are limited in their ability to adapt to dynamic loads. Programmable thermal materials can direct heat away from hotspots in real time, maintain optimal operating temperatures, and even perform thermal logic operations. Microelectromechanical systems can act as thermal switches, selectively connecting or disconnecting thermal pathways, while phase-change elements can provide reversible insulation or conduction. These advances enable high-performance computing, flexible electronics, and compact wearable devices that can operate safely and efficiently under variable thermal conditions.

6.3.5 Energy Harvesting and Thermoregulation

Programmable thermal behavior also enhances energy harvesting and thermoregulation systems. Thermoelectric devices, which convert temperature gradients into electrical energy, can benefit from programmable heat flow to optimize the gradient across the device. Similarly, in solar-thermal systems, redirecting heat dynamically to specific regions enhances energy capture efficiency. Programmable surfaces can also maintain thermal comfort in buildings by modulating emissivity or insulation properties according to environmental conditions. For example, a wall coated with a programmable thermal material might reflect heat during daytime and retain warmth at night, reducing energy consumption for heating and cooling. In biomedical applications, programmable thermal patches could maintain localized temperatures for therapeutic purposes or regulate heat dissipation in implanted devices.

6.3.6 Thermal Logic and Computing

Beyond conventional thermal management, programmable thermal behavior enables novel computational paradigms using heat as a medium for information processing. Thermal logic exploits the fact that heat fluxes can be guided, modulated, and switched, allowing devices to perform logical operations similar to electronic circuits. By integrating thermal diodes, transistors, and switches made from phase-change or anisotropic materials, it is possible to construct thermal circuits that process information through temperature differences. Programmable control enables these circuits to adapt, perform multiple functions, or respond dynamically to inputs. Such thermal computing systems could complement electronic circuits in high-temperature or radiation-intense environments where conventional electronics may fail.

6.3.7 Defense and Stealth Applications

Programmable thermal behavior has clear implications for defense and stealth technologies. Infrared detection systems are widely used for surveillance, targeting, and threat assessment. By dynamically controlling thermal signatures, programmable metamaterials can hide or mimic thermal profiles, creating adaptive thermal camouflage. Active thermal cloaks can redirect heat around critical components, rendering them invisible to infrared imaging. In addition, programmable emissivity surfaces can simulate environmental backgrounds, further enhancing stealth capabilities. These applications demonstrate how dynamic control over heat can provide strategic advantages in military and security operations.

6.3.8 Challenges in Realization

While the potential of programmable thermal behavior is immense, several challenges must be addressed to realize its full potential. One major difficulty is material limitations, including the range of thermal conductivity modulation and the speed of response. Phase-change materials, for instance, can provide significant contrast in thermal properties but may suffer from fatigue after repeated cycling. Integration with electronic or optical controls also requires careful design to minimize energy consumption and maximize reliability. Fabrication complexity is another challenge, as creating micro- or nanoscale metamaterials with active components demands precision and scalability. Furthermore, ensuring stable performance under varying environmental conditions, such as temperature extremes or mechanical stress, remains a critical concern.

6.4 THERMAL METASURFACES FOR ENERGY HARVESTING

6.4.1 Thermal Metasurfaces

Thermal metasurfaces are engineered surfaces with subwavelength patterns designed to control thermal radiation and heat flow. These structures exploit the principles of metamaterials, where the overall thermal behavior is determined by geometry rather than the base material alone.

By manipulating thermal emission, reflection, and absorption, metasurfaces can achieve phenomena such as directional radiative heat transfer, selective absorption, and wavelength-specific emission. The programmable nature of these metasurfaces allows dynamic control over thermal behavior, enabling real-time adaptation to changing environmental conditions. In energy harvesting, this capability is particularly valuable because it maximizes the extraction of thermal energy from ambient or waste heat sources.

6.4.2 Mechanisms for Heat Control

Thermal metasurfaces achieve control over energy flow through several mechanisms. One approach involves structuring the surface at the nanoscale to create resonant cavities that trap and absorb thermal radiation efficiently. Another method uses gradient-index patterns to redirect heat flux along desired pathways, effectively concentrating energy in specific regions. Phase-change materials incorporated into metasurfaces enable dynamic adjustment of thermal conductivity and emissivity, allowing the surface to switch between high absorption and low emission states. Additionally, plasmonic structures can enhance local electromagnetic fields, increasing radiative heat transfer at targeted wavelengths. These mechanisms collectively provide the precision necessary for efficient thermal energy harvesting.

6.4.3 Integration with Thermoelectric Devices

A primary application of thermal metasurfaces is in thermoelectric energy harvesting, where temperature gradients are converted into electrical energy. Conventional thermoelectric materials often suffer from low efficiency due to limited control over heat flow and uneven temperature distribution. By integrating thermal metasurfaces, heat can be directed and concentrated onto thermoelectric elements, enhancing the local temperature gradient and boosting power output. Nanoscale surface patterns can also suppress parasitic heat loss, ensuring that more energy contributes to electrical generation. In addition, metasurfaces can be tuned to harvest energy across different thermal spectra, making them highly adaptable for various heat sources, including industrial waste heat, solar radiation, and environmental temperature differences.

6.4.4 Solar-Thermal Energy Applications

Thermal metasurfaces are particularly valuable in solar-thermal energy systems. Traditional solar collectors often face limitations in absorption efficiency and spectral selectivity. Metasurfaces can be engineered to selectively absorb sunlight in the visible and near-infrared range while minimizing thermal radiation losses in the mid-infrared. This spectral control increases the effective thermal energy captured by the system, improving overall efficiency. Furthermore, metasurfaces can concentrate heat in small regions, reducing the amount of active material needed for conversion and allowing for compact, high-performance solar-thermal devices. By integrating phase-change elements or tunable materials, metasurfaces can dynamically adjust absorption based on incident sunlight intensity, providing adaptive performance throughout the day.

6.4.5 Waste Heat Recovery

Beyond solar energy, thermal metasurfaces have significant potential in waste heat recovery. Industrial processes, power plants, and electronic devices generate large amounts of heat that is often lost to the environment. Metasurfaces can be employed to redirect and concentrate this heat toward energy conversion systems such as thermoelectric generators. By optimizing surface geometry and material properties, metasurfaces enhance the capture of thermal energy that would otherwise dissipate. In microelectronics, metasurfaces can direct heat from hotspots toward thermoelectric layers, converting excess heat into usable electricity while simultaneously improving device cooling. Such applications contribute to energy efficiency and sustainability in both industrial and consumer technologies.

6.4.6 Dynamic and Tunable Metasurfaces

One of the most exciting aspects of thermal metasurfaces for energy harvesting is their tunability. By incorporating active materials such as vanadium dioxide, graphene, or liquid crystals, metasurfaces can adjust their thermal properties in response to external stimuli. This dynamic control allows for real-time optimization of heat absorption, concentration, and conversion efficiency. For example, a metasurface integrated with a phase-change material can switch between high and low thermal conductivity states, adapting to fluctuating heat sources. Graphene layers can modulate thermal emission under applied voltage, enabling adaptive control of radiative heat transfer. This programmability ensures that energy harvesting systems can maintain high performance across varying environmental and operational conditions.

6.4.7 Materials and Fabrication Techniques

The effectiveness of thermal metasurfaces depends heavily on material selection and fabrication precision. High-conductivity metals, dielectric coatings, and phase-change compounds are commonly used to achieve desired thermal behaviors. Nanoscale patterning techniques such as electron-beam lithography, nanoimprinting, and focused ion beam milling allow precise control over surface features, ensuring optimal interaction with thermal radiation. Recent advances in 3D printing and additive manufacturing have enabled the production of complex metasurface geometries at larger scales, paving the way for practical deployment in energy harvesting systems. Material innovations, such as hybrid composites combining plasmonic and dielectric elements, further expand the design space and efficiency of metasurfaces.

6.4.8 Challenges in Implementation

Despite significant progress, challenges remain in the development of thermal metasurfaces for energy harvesting. Fabrication at the nanoscale can be costly and difficult to scale for large-area applications. Achieving durability and stability under prolonged thermal cycling is also critical, particularly for outdoor or industrial environments. Integrating metasurfaces with existing energy conversion systems requires careful design to minimize thermal and mechanical mismatches. Additionally, optimizing performance across broad temperature ranges and multiple spectral regions remains a complex challenge. Addressing these issues is essential to realize the full potential of metasurfaces in practical, real-world energy harvesting applications.

CHAPTER 7

Design, Simulation, and Fabrication

7.1 COMPUTATIONAL DESIGN TOOLS

7.1.1 FDTD

Finite-difference time-domain (FDTD) or Yee's method (named after the Chinese American applied mathematician Kane S. Yee, born 1934) is a numerical analysis technique used for modeling computational electrodynamics.

History

Finite difference schemes for time-dependent partial differential equations (PDEs) have been employed for many years in computational fluid dynamics problems, including the idea of using centered finite difference operators on staggered grids in space and time to achieve second-order accuracy. The novelty of Yee's FDTD scheme, presented in his seminal 1966 paper, was to apply centered finite difference operators on staggered grids in space and time for each electric and magnetic vector field component in Maxwell's curl equations. The descriptor "Finite-difference time-domain" and its corresponding "FDTD" acronym were originated by Allen Taflove in 1980. Since about 1990, FDTD techniques have emerged as primary means to computationally model many scientific and engineering problems dealing with electromagnetic wave interactions with material structures.

Current FDTD modeling applications range from near-DC (ultralow-frequency geophysics involving the entire Earth-ionosphere waveguide) through microwaves (radar signature technology, antennas, wireless communications devices, digital interconnects, biomedical imaging/treatment) to visible light (photonic crystals, nanoplasmonics, solitons, and biophotonics). In 2006, an estimated 2,000 FDTD-related publications appeared in the science and engineering literature. As of 2013, there are at least 25 commercial/proprietary FDTD software vendors; 13 free-software/open-source-software FDTD projects; and 2 freeware/closed-source FDTD projects, some not for commercial use.

Development of FDTD and Maxwell's Equations

An appreciation of the basis, technical development, and possible future of FDTD numerical techniques for Maxwell's equations can be developed by first considering their history. The following lists some of the key publications in this area.

FDTD Models and Methods

When Maxwell's differential equations are examined, it can be seen that the change in the E-field in time (the time derivative) is dependent on the change in the H-field across space (the curl). This results in the basic FDTD time-stepping relation that, at any point in space, the updated value of the E-field in time is dependent on the stored value of the E-field and the numerical curl of the local distribution of the H-field in space. The H-field is time-stepped in a similar manner. At any point in space, the updated value of the H-field in time is dependent on the stored value of the H-field and the numerical curl of the local distribution of the E-field in space. Iterating the E-field and H-field updates results in a marching-in-time process wherein sampled-data analogs of the continuous electromagnetic waves under consideration propagate in a numerical grid stored in the computer memory.

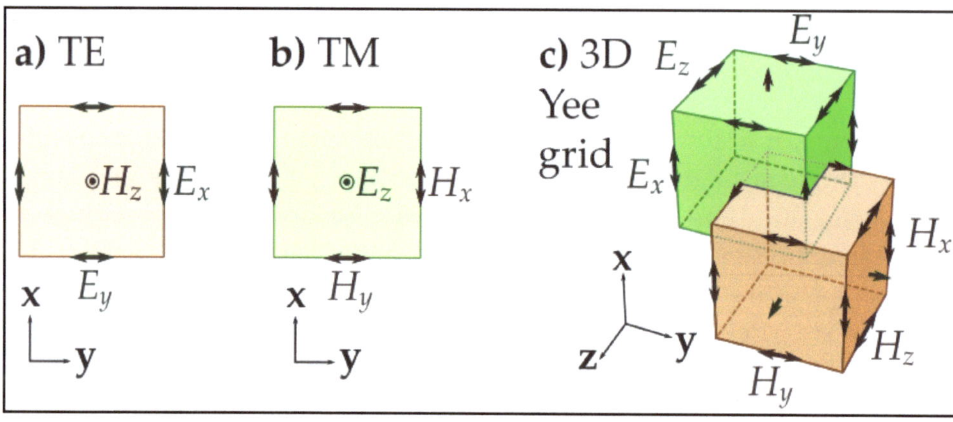

Illustration of a Standard Cartesian Yee Cell used for FDTD, about which Electric and Magnetic Field Vector Components are Distributed

Visualized as a cubic voxel, the electric field components form the edges of the cube, and the magnetic field components form the normals to the faces of the cube. A three-dimensional space lattice consists of a multiplicity of such Yee cells. An electromagnetic wave interaction structure is mapped into the space lattice by assigning appropriate values of permittivity to each electric field component, and permeability to each magnetic field component.

This description holds true for 1-D, 2-D, and 3-D FDTD techniques. When multiple dimensions are considered, calculating the numerical curl can become complicated. Kane Yee's seminal 1966 paper proposed spatially staggering the vector components of the E-field and H-field about rectangular unit cells of a Cartesian computational grid so that each E-field vector component is located midway between a pair of H-field vector components, and conversely. This scheme, now known as a Yee lattice, has proven to be very robust, and remains at the core of many current FDTD software constructs. Furthermore, Yee proposed a leapfrog scheme for marching in time wherein the E-field and H-field updates are staggered so that E-field updates are conducted midway during each time-step between successive H-field updates, and conversely. On the plus side, this explicit time-stepping scheme avoids the need to solve simultaneous equations, and furthermore yields dissipation-free numerical wave propagation. On the minus side, this scheme mandates an upper bound on the time-step to ensure numerical stability. As a result, certain classes of simulations can require many thousands of time-steps for completion.

Using the FDTD Method

To implement an FDTD solution of Maxwell's equations, a computational domain must first be established. The computational domain is simply the physical region over which the simulation will be performed. The E and H fields are determined at every point in space within that computational domain. The material of each cell within the computational domain must be specified. Typically, the material is either free-space (air), metal, or dielectric. Any material can be used as long as the permeability, permittivity, and conductivity are specified.

The permittivity of dispersive materials in tabular form cannot be directly substituted into the FDTD scheme. Instead, it can be approximated using multiple Debye, Drude, Lorentz or critical point terms. This approximation can be obtained using open fitting programs and does not necessarily have physical meaning. Once the computational domain and the grid materials are established, a source is specified. The source can be current on a wire, applied electric field or impinging plane wave. In the last case FDTD can be used to simulate light scattering from arbitrary shaped objects, planar periodic structures at various incident angles, and photonic band structure of infinite periodic structures. Since the E and H fields are determined directly, the output of the simulation is usually the E or H field at a point or a series of points within the computational domain. The simulation evolves the E and H fields forward in time. Processing may be done on the E and H fields returned by the simulation. Data processing may also occur while the simulation is ongoing. While the FDTD technique computes electromagnetic fields within a compact spatial region, scattered and/or radiated far fields can be obtained via near-to-far-field transformations.

- **Stability:** Due to the linearity of the FDTD method, the region of stability of the FDTD method may be determined by Von Neumann stability analysis. This method assumes that electric and magnetic fields are proportional to a monochromatic complex exponential. After a single time-step, the magnitude amplitude of the stable fields need to remain the same or less. This leads to the Courant–Friedrichs–Lewy condition, which describes the relationship of the FDTD parameters to ensure stability.

Strengths of FDTD Modeling

Every modeling technique has strengths and weaknesses, and the FDTD method is no different.

- FDTD is a versatile modeling technique used to solve Maxwell's equations. It is intuitive, so users can easily understand how to use it and know what to expect from a given model.
- FDTD is a time-domain technique, and when a broadband pulse (such as a Gaussian pulse) is used as the source, then the response of the system over a wide range of frequencies can be obtained with a single simulation. This is useful in applications where resonant frequencies are not exactly known, or anytime that a broadband result is desired.
- Since FDTD calculates the E and H fields everywhere in the computational domain as they evolve in time, it lends itself to providing animated displays of the electromagnetic field movement through the model. This type of display is useful in understanding what is going on in the model, and to help ensure that the model is working correctly.
- The FDTD technique allows the user to specify the material at all points within the computational domain. A wide variety of linear and nonlinear dielectric and magnetic materials can be naturally and easily modeled.
- FDTD allows the effects of apertures to be determined directly. Shielding effects can be found, and the fields both inside and outside a structure can be found directly or indirectly.
- FDTD uses the E and H fields directly. Since most EMI/EMC modeling applications are interested in the E and H fields, it is convenient that no conversions must be made after the simulation has run to get these values.

Weaknesses of FDTD Modeling

- Since FDTD requires that the entire computational domain be gridded, and the grid spatial discretization must be sufficiently fine to resolve both the smallest electromagnetic wavelength and the smallest geometrical feature in the model, very large computational domains can be developed, which results in very long solution times. Models with long, thin features, (like wires) are difficult to model in FDTD because of the excessively large computational domain required. Methods such as eigenmode expansion can offer a more efficient alternative as they do not require a fine grid along the z-direction.

- There is no way to determine unique values for permittivity and permeability at a material interface.
- Space and time steps must satisfy the CFL condition, or the leapfrog integration used to solve the partial differential equation is likely to become unstable.
- FDTD finds the E/H fields directly everywhere in the computational domain. If the field values at some distance are desired, it is likely that this distance will force the computational domain to be excessively large. Far-field extensions are available for FDTD, but require some amount of postprocessing.
- Since FDTD simulations calculate the E and H fields at all points within the computational domain, the computational domain must be finite to permit its residence in the computer memory. In many cases this is achieved by inserting artificial boundaries into the simulation space. Care must be taken to minimize errors introduced by such boundaries. There are a number of available highly effective absorbing boundary conditions (ABCs) to simulate an infinite unbounded computational domain. Most modern FDTD implementations instead use a special absorbing "material", called a perfectly matched layer (PML) to implement absorbing boundaries.
- Because FDTD is solved by propagating the fields forward in the time domain, the electromagnetic time response of the medium must be modeled explicitly. For an arbitrary response, this involves a computationally expensive time convolution, although in most cases the time response of the medium (or Dispersion (optics)) can be adequately and simply modeled using either the recursive convolution (RC) technique, the auxiliary differential equation (ADE) technique, or the Z-transform technique. An alternative way of solving Maxwell's equations that can treat arbitrary dispersion easily is the pseudo-spectral spatial domain (PSSD), which instead propagates the fields forward in space.

Grid Truncation Techniques

The most commonly used grid truncation techniques for open-region FDTD modeling problems are the Mur absorbing boundary condition (ABC), the Liao ABC, and various perfectly matched layer (PML) formulations. The Mur and Liao techniques are simpler than PML. However, PML (which is technically an absorbing region rather than a boundary condition per se) can provide orders-of-magnitude lower reflections. The PML concept was introduced by J.-P. Berenger in a seminal

1994 paper in the Journal of Computational Physics. Since 1994, Berenger's original split-field implementation has been modified and extended to the uniaxial PML (UPML), the convolutional PML (CPML), and the higher-order PML. The latter two PML formulations have increased ability to absorb evanescent waves, and therefore can in principle be placed closer to a simulated scattering or radiating structure than Berenger's original formulation. To reduce undesired numerical reflection from the PML additional back absorbing layers technique can be used.

Popularity

Notwithstanding both the general increase in academic publication throughput during the same period and the overall expansion of interest in all Computational electromagnetics (CEM) techniques, there are seven primary reasons for the tremendous expansion of interest in FDTD computational solution approaches for Maxwell's equations:

- FDTD does not require a matrix inversion. Being a fully explicit computation, FDTD avoids the difficulties with matrix inversions that limit the size of frequency-domain integral-equation and finite-element electromagnetics models to generally fewer than 10^9 electromagnetic field unknowns. FDTD models with as many as 10^9 field unknowns have been run; there is no intrinsic upper bound to this number.
- FDTD is accurate and robust. The sources of error in FDTD calculations are well understood, and can be bounded to permit accurate models for a very large variety of electromagnetic wave interaction problems.
- FDTD treats impulsive behavior naturally. Being a time-domain technique, FDTD directly calculates the impulse response of an electromagnetic system. Therefore, a single FDTD simulation can provide either ultrawideband temporal waveforms or the sinusoidal steady-state response at any frequency within the excitation spectrum.
- FDTD treats nonlinear behavior naturally. Being a time-domain technique, FDTD directly calculates the nonlinear response of an electromagnetic system. This allows natural hybriding of FDTD with sets of auxiliary differential equations that describe nonlinearities from either the classical or semi-classical standpoint. One research frontier is the development of hybrid algorithms which join FDTD classical electrodynamics models with phenomena arising from quantum electrodynamics, especially vacuum fluctuations, such as the Casimir effect.
- FDTD is a systematic approach. With FDTD, specifying a new structure to be modeled is reduced to a problem of mesh generation rather than

the potentially complex reformulation of an integral equation. For example, FDTD requires no calculation of structure-dependent Green functions.

- Parallel-processing computer architectures have come to dominate supercomputing. FDTD scales with high efficiency on parallel-processing CPU-based computers, and extremely well on recently developed GPU-based accelerator technology.
- Computer visualization capabilities are increasing rapidly. While this trend positively influences all numerical techniques, it is of particular advantage to FDTD methods, which generate time-marched arrays of field quantities suitable for use in color videos to illustrate the field dynamics.
- Anisotropy is treated naturally by the FDTD method. Yee cells, having components in each Cartesian direction, can be easily configured with anisotropic characteristics.

7.1.2 FEM

Basic Concepts

FEM Mesh Created by an Analyst before Finding a Solution to a Magnetic Problem using FEM software

Finite element method (FEM) is a popular method for numerically solving differential equations arising in engineering and mathematical modeling. Typical problem areas of interest include the traditional fields of structural analysis, heat transfer, fluid flow, mass transport, and electromagnetic potential. Computers are usually used to perform the calculations required. With high-speed supercomputers, better solutions can be achieved and are often required to solve the largest and most complex problems.

FEM is a general numerical method for solving partial differential equations in two- or three-space variables (i.e., some boundary value problems). There are also studies about using FEM to solve high-dimensional problems. To solve a problem, FEM subdivides a large system into smaller, simpler parts called finite elements. This is achieved by a particular space discretization in the space dimensions, which is implemented by the construction of a mesh of the object: the numerical domain for the solution that has a finite number of points. FEM formulation of a boundary value problem finally results in a system of algebraic equations. The method approximates the unknown function over the domain. The simple equations that model these finite elements are then assembled into a larger system of equations that models the entire problem. FEM then approximates a solution by minimizing an associated error function via the calculus of variations. Studying or analyzing a phenomenon with FEM is often referred to as finite element analysis (FEA).

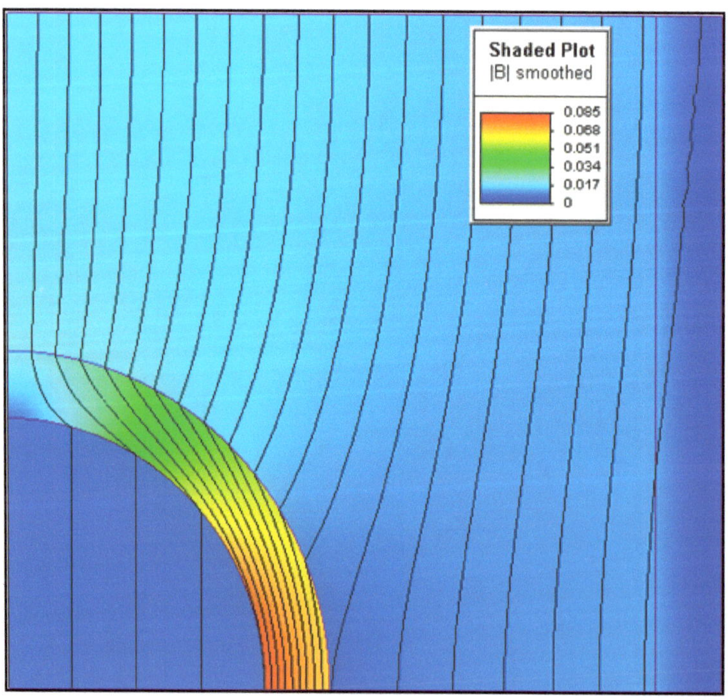

FEM Solution to the Problem at Left, Involving a Cylindrically Shaped Magnetic Shield

Colors indicate that the analyst has set material properties for each zone, in this case, a conducting wire coil in orange; a ferromagnetic component (perhaps iron) in light blue; and air in grey. Although the geometry may seem simple, it would be very challenging to calculate the magnetic field for this setup without FEM software using equations alone. The ferromagnetic cylindrical part shields the area inside the cylinder by diverting the magnetic field created by the coil (rectangular area on the right). The color represents the amplitude of the magnetic flux density, as indicated by the scale in the inset legend, red being high amplitude. The area inside the cylinder is low amplitude (dark blue, with widely spaced lines of magnetic flux), which suggests that the shield is performing as it was designed to.

The subdivision of a whole domain into simpler parts has several advantages:

- Accurate representation of complex geometry;
- Inclusion of dissimilar material properties;
- Easy representation of the total solution; and
- Capture of local effects.

A typical approach using the method involves the following steps:

- Dividing the domain of the problem into a collection of subdomains, with each subdomain represented by a set of element equations for the original problem.
- Systematically recombining all sets of element equations into a global system of equations for the final calculation.

The global system of equations uses known solution techniques and can be calculated from the initial values of the original problem to obtain a numerical answer.

The element equations are simple equations that locally approximate the original complex equations to be studied, where the original equations are often partial differential equations (PDEs). To explain the approximation of this process, FEM is commonly introduced as a special case of the Galerkin method. The process, in mathematical language, is to construct an integral of the inner product of the residual and the weight functions; then, set the integral to zero. In simple terms, it is a procedure that minimizes the approximation error by fitting trial functions into the PDE. The residual is the error caused by the trial functions, and the weight functions are polynomial approximation functions that project the residual. The process eliminates all the spatial derivatives from the PDE, thus approximating the PDE locally using the following:

- a set of algebraic equations for steady-state problems; and
- a set of ordinary differential equations for transient problems.

These equation sets are element equations. They are linear if the underlying PDE is linear and vice versa. Algebraic equation sets that arise in the steady-state problems are solved using numerical linear algebraic methods. In contrast, ordinary differential equation sets that occur in the transient problems are solved by numerical integrations using standard techniques such as Euler's method or the Runge–Kutta method.

A global system of equations is generated from the element equations by transforming coordinates from the subdomains' local nodes to the domain's global nodes. This spatial transformation includes appropriate orientation adjustments as applied in relation to the reference coordinate system. The process is often carried out using FEM software with coordinate data generated from the subdomains.

The practical application of FEM is known as finite element analysis (FEA). FEA, as applied in engineering, is a computational tool for performing engineering analysis. It includes the use of mesh generation techniques for dividing a complex problem into smaller elements, as well as the use of software coded with a FEM algorithm. When applying FEA, the complex problem is usually a physical system with the underlying physics, such as the Euler–Bernoulli beam equation, the heat equation, or the Navier–Stokes equations, expressed in either PDEs or integral equations, while the divided, smaller elements of the complex problem represent different areas in the physical system.

FEA may be used for analyzing problems over complicated domains (e.g., cars and oil pipelines) when the domain changes (e.g., during a solid-state reaction with a moving boundary), when the desired precision varies over the entire domain, or when the solution lacks smoothness. FEA simulations provide a valuable resource, as they remove multiple instances of creating and testing complex prototypes for various high-fidelity situations.

For example, in a frontal crash simulation, it is possible to increase prediction accuracy in important areas, like the front of the car, and reduce it in the rear of the car, thus reducing the cost of the simulation. Another example would be in numerical weather prediction, where it is more important to have accurate predictions over developing highly nonlinear phenomena, such as tropical cyclones in the atmosphere or eddies in the ocean, rather than relatively calm areas. A clear, detailed, and practical presentation of this approach can be found in the textbook The Finite Element Method for Engineers.

History

While it is difficult to quote the date of the invention of FEM, the method originated from the need to solve complex elasticity and structural analysis problems in civil and aeronautical engineering. Its development can be traced back to work by

Alexander Hrennikoff and Richard Courant in the early 1940s. Another pioneer was Ioannis Argyris. In the USSR, the introduction of the practical application of FEM is usually connected with Leonard Oganesyan.

It was also independently rediscovered in China by Feng Kang in the late 1950s and early 1960s, based on the computations of dam constructions, where it was called the "finite difference method" based on variation principles. Although the approaches used by these pioneers are different, they share one essential characteristic: the mesh discretization of a continuous domain into a set of discrete sub-domains, usually called elements.

Hrennikoff's work discretizes the domain by using a lattice analogy, while Courant's approach divides the domain into finite triangular sub-regions to solve second-order elliptic partial differential equations that arise from the problem of the torsion of a cylinder. Courant's contribution was evolutionary, drawing on a large body of earlier results for PDEs developed by Lord Rayleigh, Walther Ritz, and Boris Galerkin.

The application of FEM gained momentum in the 1960s and 1970s due to the developments of J. H. Argyris and his co-workers at the University of Stuttgart; R. W. Clough and his co-workers at University of California Berkeley; O. C. Zienkiewicz and his co-workers Ernest Hinton, Bruce Irons, and others at Swansea University; Philippe G. Ciarlet at the University of Paris 6; and Richard Gallagher and his co-workers at Cornell University. During this period, additional impetus was provided by the available open-source FEM programs.

NASA sponsored the original version of NASTRAN. University of California Berkeley made the finite element programs SAP IV and, later, OpenSees widely available. In Norway, the ship classification society Det Norske Veritas (now DNV GL) developed Sesam in 1969 for use in the analysis of ships.

A rigorous mathematical basis for FEM was provided in 1973 with a publication by Gilbert Strang and George Fix. The method has since been generalized for the numerical modeling of physical systems in a wide variety of engineering disciplines, such as electromagnetism, heat transfer, and fluid dynamics.

Technical Discussion

The Structure of Finite Element Methods

A finite element method is characterized by a variational formulation, a discretization strategy, one or more solution algorithms, and post-processing procedures. Examples of the variational formulation are the Galerkin method, the discontinuous Galerkin method, mixed methods, etc. A discretization strategy is understood to mean a clearly defined set of procedures that cover:

- The creation of finite element meshes,
- The definition of basis function on reference elements (also called shape functions), and
- the mapping of reference elements onto the elements of the mesh.

Examples of discretization strategies are the h-version, p-version, hp-version, x-FEM, isogeometric analysis, etc. Each discretization strategy has certain advantages and disadvantages. A reasonable criterion in selecting a discretization strategy is to realize nearly optimal performance for the broadest set of mathematical models in a particular model class. Various numerical solution algorithms can be classified into two broad categories; direct and iterative solvers. These algorithms are designed to exploit the sparsity of matrices that depend on the variational formulation and discretization strategy choices. Post-processing procedures are designed to extract the data of interest from a finite element solution. To meet the requirements of solution verification, postprocessors need to provide for a posteriori error estimation in terms of the quantities of interest. When the errors of approximation are larger than what is considered acceptable, then the discretization has to be changed either by an automated adaptive process or by the action of the analyst. Some very efficient postprocessors provide for the realization of superconvergence.

Illustrative Problems P1 and P2

The following two problems demonstrate the finite element method.

P1 is a one-dimensional problem

$$\mathrm{P1}\ :\begin{cases} u''(x) = f(x) \text{ in } (0,1), \\ u(0) = u(1) = 0, \end{cases}$$

P2 is a two-dimensional problem (Dirichlet problem) P2 :

$$\mathrm{P2}\ :\begin{cases} u_{xx}(x,y) + u_{yy}(x,y) = f(x,y) & \text{in } \Omega, \\ u = 0 & \text{on } \partial\Omega, \end{cases}$$

The problem P1 can be solved directly by computing antiderivatives. However, this method of solving the boundary value problem (BVP) works only when there is one spatial dimension. It does not generalize to higher-dimensional problems or problems like $u + V'' = f$. For this reason, we will develop the finite element method for P1 and outline its generalization to P2. Our explanation will proceed in two steps, which mirror two essential steps one must take to solve a boundary value problem (BVP) using the FEM.

- In the first step, one rephrases the original BVP in its weak form. Little to no computation is usually required for this step. The transformation is done by hand on paper.

- The second step is discretization, where the weak form is discretized in a finite-dimensional space.

After this second step, we have concrete formulae for a large but finite-dimensional linear problem whose solution will approximately solve the original BVP. This finite-dimensional problem is then implemented on a computer.

Weak Formulation

The first step is to convert P1 and P2 into their equivalent weak formulations.

The Weak form of P1

If u solves P1, then for any smooth function v that satisfies the displacement boundary conditions, i.e. $v = 0$ at $x = 0$ and $x = 1$, we have

$$\int_0^1 f(x)v(x)\,dx = \int_0^1 u''(x)v(x)\,dx.$$

Conversely, if u with $u(0) = u(1) = 0$ satisfies (1) for every smooth function $v(x)$ then one may show that this u will solve P1. The proof is easier for twice continuously differentiable u (mean value theorem) but may be proved in a distributional sense as well. We define a new operator or map $\phi(u, v)$ by using integration by parts on the right-hand-side of (1):

$$\begin{aligned}\int_0^1 f(x)v(x)dx &= \int_0^1 u''(x)v(x)dx \\ &= u'(x)v(x)\,|_0^1 - \int_0^1 u'(x)v'(x)dx \\ &= -\int_0^1 u'(x)v'(x)dx \equiv -\phi(u, v),\end{aligned}$$

The Weak form of P2

If we integrate by parts using a form of Green's identities, we see that if u u solves P2, then we may define $\phi(u, v)$ for any v by

$$\int_\Omega fv\,ds = -\int_\Omega \nabla u \cdot \nabla v\,ds \equiv -\phi(u, v),$$

A Proof Outline of the Existence and Uniqueness of the Solution

We can loosely think of $H_0^1(0,1)$ to be the absolutely continuous functions of $(0,1)$ that are 0 at $x = 0$ and $x = 1$. Such functions are (weakly) once differentiable, and it turns out that the symmetric bilinear map ϕ then defines an inner product which turns $H_0^1(0,1)$ into a Hilbert space (a detailed proof is nontrivial). On the other hand, the left-hand-side $\int_0^1 f(x)v(x)dx$ is also an inner product, this time on the Lp space

$L^2(0,1)$. An application of the Riesz representation theorem for Hilbert spaces shows that there is a unique u solving (2) and, therefore, P1. This solution is a-priori only a member of $H_0^1(0,1)$, but using elliptic regularity, will be smooth if f is.

Discretization

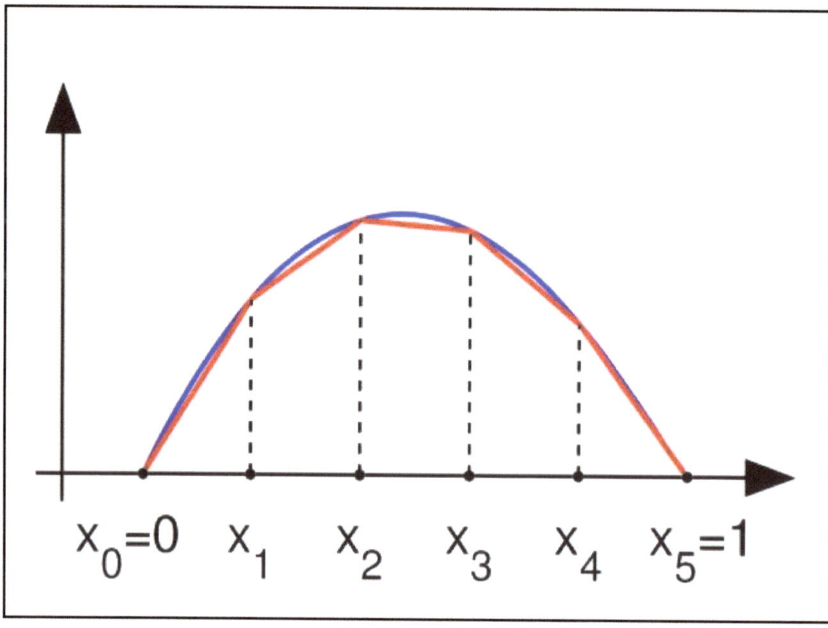

A Function in with Zero Values at the Endpoints (Blue) and a Piecewise Linear Approximation (Red)

P1 and P2 are ready to be discretized, which leads to a common sub-problem (3). The basic idea is to replace the infinite-dimensional linear problem:

Find $u \in H_0^1$ such that

$$\forall v \in H_0^1, -\phi(u,v) = \int fv$$

with a finite-dimensional version:

Find $u \in V$ such that

$$\forall v \in V, -\phi(u,v) = \int fv$$

Various Types of Finite Element Methods

- **AEM:** The Applied Element Method or AEM combines features of both FEM and Discrete element method or (DEM).
- **A-FEM:** Yang and Lui introduced the Augmented-Finite Element Method, whose goal was to model the weak and strong discontinuities without needing extra DoFs, as PuM stated.

- **CutFEM:** The Cut Finite Element Approach was developed in 2014. The approach is "to make the discretization as independent as possible of the geometric description and minimize the complexity of mesh generation, while retaining the accuracy and robustness of a standard finite element method."
- **Generalized Finite Element Method:** The generalized finite element method (GFEM) uses local spaces consisting of functions, not necessarily polynomials, that reflect the available information on the unknown solution and thus ensure good local approximation. Then a partition of unity is used to "bond" these spaces together to form the approximating subspace. The effectiveness of GFEM has been shown when applied to problems with domains having complicated boundaries, problems with micro-scales, and problems with boundary layers.
- **Mixed Finite Element Method:** The mixed finite element method is a type of finite element method in which extra independent variables are introduced as nodal variables during the discretization of a partial differential equation problem.
- **Variable—Polynomial:** The hp-FEM combines adaptively elements with variable size h and polynomial degree p to achieve exceptionally fast, exponential convergence rates.
- **Hpk-FEM:** The hpk-FEM combines adaptively elements with variable size h, polynomial degree of the local approximations p, and global differentiability of the local approximations (k-1) to achieve the best convergence rates.
- **XFEM:** The extended finite element method (XFEM) is a numerical technique based on the generalized finite element method (GFEM) and the partition of unity method (PUM). It extends the classical finite element method by enriching the solution space for solutions to differential equations with discontinuous functions. Extended finite element methods enrich the approximation space to naturally reproduce the challenging feature associated with the problem of interest: the discontinuity, singularity, boundary layer, etc. It was shown that for some problems, such an embedding of the problem's feature into the approximation space can significantly improve convergence rates and accuracy. Moreover, treating problems with discontinuities with XFEMs suppresses the need to mesh and re-mesh the discontinuity surfaces, thus alleviating the computational costs and projection errors associated with conventional finite element methods at the cost of

restricting the discontinuities to mesh edges. Several research codes implement this technique to various degrees:

- GetFEM++
- xfem++
- openxfem++

- **Scaled Boundary Finite Element Method (SBFEM):** The introduction of the scaled boundary finite element method (SBFEM) came from Song and Wolf (1997). The SBFEM has been one of the most profitable contributions in the area of numerical analysis of fracture mechanics problems. It is a semi-analytical fundamental-solutionless method combining the advantages of finite element formulations and procedures and boundary element discretization. However, unlike the boundary element method, no fundamental differential solution is required.
- **S-FEM:** The S-FEM, Smoothed Finite Element Methods, is a particular class of numerical simulation algorithms for the simulation of physical phenomena. It was developed by combining mesh-free methods with the finite element method.
- **Spectral Element Method:** Spectral element methods combine the geometric flexibility of finite elements and the acute accuracy of spectral methods. Spectral methods are the approximate solution of weak-form partial equations based on high-order Lagrangian interpolants and used only with certain quadrature rules.
- **Crystal Plasticity Finite Element Method (CPFEM):** The crystal plasticity finite element method (CPFEM) is an advanced numerical tool developed by Franz Roters. Metals can be regarded as crystal aggregates, which behave anisotropy under deformation, such as abnormal stress and strain localization. CPFEM, based on the slip (shear strain rate), can calculate dislocation, crystal orientation, and other texture information to consider crystal anisotropy during the routine. It has been applied in the numerical study of material deformation, surface roughness, fractures, etc.
- **Virtual Element Method (VEM):** The virtual element method (VEM), introduced by Beirão da Veiga et al. (2013) as an extension of mimetic finite difference (MFD) methods, is a generalization of the standard finite element method for arbitrary element geometries. This allows admission of general polygons (or polyhedra in 3D) that are highly irregular and non-convex in shape. The name virtual derives from the fact that knowledge of the local shape function basis is not required and is, in fact, never explicitly calculated.

Link with the Gradient Discretization Method

Some types of finite element methods (conforming, nonconforming, mixed finite element methods) are particular cases of the gradient discretization method (GDM). Hence the convergence properties of the GDM, which are established for a series of problems (linear and nonlinear elliptic problems, linear, nonlinear, and degenerate parabolic problems), hold as well for these particular FEMs.

Comparison to the Finite Difference Method

The finite difference method (FDM) is an alternative way of approximating solutions of PDEs. The differences between FEM and FDM are:

- The most attractive feature of the FEM is its ability to handle complicated geometries (and boundaries) with relative ease. While FDM in its basic form is restricted to handle rectangular shapes and simple alterations thereof, the handling of geometries in FEM is theoretically straightforward.
- FDM is not usually used for irregular CAD geometries but more often for rectangular or block-shaped models.
- FEM generally allows for more flexible mesh adaptivity than FDM.
- The most attractive feature of finite differences is that it is straightforward to implement.
- One could consider the FDM a particular case of the FEM approach in several ways. E.g., first-order FEM is identical to FDM for Poisson's equation if the problem is discretized by a regular rectangular mesh with each rectangle divided into two triangles.
- There are reasons to consider the mathematical foundation of the finite element approximation more sound, for instance, because the quality of the approximation between grid points is poor in FDM.
- The quality of a FEM approximation is often higher than in the corresponding FDM approach, but this is highly problem-dependent, and several examples to the contrary can be provided.

Generally, FEM is the method of choice in all types of analysis in structural mechanics (i.e., solving for deformation and stresses in solid bodies or dynamics of structures). In contrast, computational fluid dynamics (CFD) tend to use FDM or other methods like finite volume method (FVM). CFD problems usually require discretization of the problem into a large number of cells/gridpoints (millions and more). Therefore the cost of the solution favors simpler, lower-order approximation within each cell. This is especially true for 'external flow' problems, like airflow around the car, airplane, or weather simulation.

Finite Element and Fast Fourier Transform (FFT) Methods

Another method used for approximating solutions to a partial differential equation is the Fast Fourier Transform (FFT), where the solution is approximated by a fourier series computed using the FFT. For approximating the mechanical response of materials under stress, FFT is often much faster, but FEM may be more accurate. One example of the respective advantages of the two methods is in simulation of rolling a sheet of aluminum (an FCC metal), and drawing a wire of tungsten (a BCC metal). This simulation did not have a sophisticated shape update algorithm for the FFT method. In both cases, the FFT method was more than 10 times as fast as FEM, but in the wire drawing simulation, where there were large deformations in grains, the FEM method was much more accurate. In the sheet rolling simulation, the results of the two methods were similar. FFT has a larger speed advantage in cases where the boundary conditions are given in the materials strain, and loses some of its efficiency in cases where the stress is used to apply the boundary conditions, as more iterations of the method are needed.

The FE and FFT methods can also be combined in a voxel based method (2) to simulate deformation in materials, where the FE method is used for the macroscale stress and deformation, and the FFT method is used on the microscale to deal with the effects of microscale on the mechanical response. Unlike FEM, FFT methods' similarities to image processing methods means that an actual image of the microstructure from a microscope can be input to the solver to get a more accurate stress response. Using a real image with FFT avoids meshing the microstructure, which would be required if using FEM simulation of the microstructure, and might be difficult. Because fourier approximations are inherently periodic, FFT can only be used in cases of periodic microstructure, but this is common in real materials. FFT can also be combined with FEM methods by using fourier components as the variational basis for approximating the fields inside an element, which can take advantage of the speed of FFT based solvers.

7.1.3 COMSOL

COMSOL Multiphysics is a finite element analyzer, solver, and simulation software package for various physics and engineering applications, especially coupled phenomena and multiphysics. The software facilitates conventional physics-based user interfaces and coupled systems of partial differential equations (PDEs). COMSOL Multiphysics provides an IDE and unified workflow for electrical, mechanical, fluid, acoustics, and chemical applications. Beside the classical problems that can be addressed with application modules, the core Multiphysics package can be used to solve PDEs in weak form. An API for Java and MATLAB can be used to control the software externally. The program also serves as an application builder for physics applications. Several modules are available for COMSOL, categorized according to the applications areas of Electrical, Mechanical, Fluid, Acoustic, Chemical, Multipurpose, and Interfacing.

7.1.4 CST

Computer Simulation Technology (CST) has emerged as one of the most influential tools in the study and development of metamaterials. Metamaterials are artificially engineered structures designed to exhibit electromagnetic properties not typically found in nature, such as negative refractive index, cloaking, and anomalous wave propagation. Designing and testing these materials using conventional experimental techniques is often time-consuming, expensive, and limited by fabrication constraints. CST provides a versatile simulation platform that allows researchers to model, analyze, and optimize metamaterial structures with high precision before physical implementation. Through CST, the complex interactions between electromagnetic waves and metamaterial structures can be understood at both macro and micro scales, significantly accelerating innovation in photonics, telecommunications, and sensor technologies.

7.2 TOPOLOGY OPTIMIZATION AND AI-DRIVEN DESIGN

7.2.1 Foundation of Topology Optimization

Topology optimization operates on a simple yet powerful premise: given a set of constraints, such as load conditions, boundary conditions, and a target volume, what is the most efficient distribution of material? Traditional design often starts with a preconceived shape that is then analyzed and modified. In contrast, topology optimization begins with a blank slate—a homogenous block of material within a defined volume—and systematically removes unnecessary elements.

Using finite element analysis and iterative algorithms, the software calculates stress distributions and identifies regions where material contributes little to the component's structural integrity. Through thousands of iterations, a complex, organic-looking structure emerges, one that often resembles the intricate forms found in nature, such as bones or tree branches, which are perfectly optimized for their load-bearing functions. The primary output is a conceptual design that maximizes performance while minimizing weight and material usage, a critical advantage in industries where every gram saved translates to significant fuel savings or performance gains.

7.2.2 Limitations and the Need for a New Paradigm

Despite its power, traditional topology optimization possesses significant limitations that create bottlenecks in the engineering workflow. The process is computationally intensive, often requiring hours or even days of processing time on high-performance computing clusters for a single design scenario. This makes exploring a wide range of design options or performing multi-objective optimization prohibitively time-consuming and expensive.

Furthermore, the resulting designs, while optimal from a structural standpoint, can be difficult or impossible to manufacture using traditional methods. They often feature complex, intricate geometries that are only feasible with additive manufacturing (3D printing), locking them into a specific production process. The process also typically operates in isolation, optimizing for a single physics—such as stiffness—without easily accounting for other critical factors like fluid dynamics, thermal management, or manufacturability constraints. These limitations created a clear need for a smarter, faster, and more holistic approach to computational design.

7.2.3 Rise of AI-Driven Design Intelligence

Artificial intelligence, particularly machine learning and deep learning, has emerged as the perfect catalyst to overcome these hurdles. AI-driven design injects a layer of predictive intelligence into the optimization loop, dramatically accelerating and enhancing the entire process. One of the most significant applications is the use of AI as a surrogate model. Instead of running computationally expensive simulations for every iteration, an AI model can be trained on a vast dataset of previous simulation results.

Once trained, this model can predict the performance of a new design almost instantaneously, reducing computation time from days to seconds. This allows engineers to explore a vastly larger design space, testing thousands of potential configurations and uncovering novel solutions that would have been missed by traditional methods. AI algorithms can also learn and incorporate complex, real-world constraints directly into the generative process, ensuring that the final output is not only structurally sound but also readily manufacturable and optimized for cost, assembly, and even aesthetics.

7.2.4 Symbiotic Workflow: From Generative Design to Realization

The integration of these two technologies creates a powerful symbiotic workflow, often referred to as generative design or AI-augmented optimization. The process begins with the engineer defining the problem: the design space, the constraints, the loads, and the performance goals. An AI-driven platform then generates a multitude of design alternatives that meet these criteria, each one a valid, optimized solution. Crucially, the AI does not produce just one answer but a palette of options, each representing a different trade-off between competing objectives, such as weight versus stiffness. The engineer can then explore this Pareto frontier, selecting the design that best balances all requirements. This transforms the designer's role from a primary creator to a curator and strategist, focusing on high-level goals and constraints while the AI handles the complex computational heavy lifting. The final selected design, often a highly complex, topology-optimized structure, is then prepared for manufacturing, frequently using additive manufacturing techniques that can realize these organic forms.

7.2.5 Transformative Applications across Industries

The impact of this AI-driven topology optimization is being felt across the entire industrial landscape. In the aerospace and automotive sectors, it is instrumental in lightweighting components, directly leading to reduced fuel consumption and lower carbon emissions. Engine brackets, wing supports, and chassis components are being redesigned with internal lattices and intricate geometries that maintain strength while shedding up to half their weight. In the medical field, it is revolutionizing the creation of patient-specific implants. By using AI to optimize a design based on a patient's own bone density and load-bearing needs, manufacturers can produce titanium hip or knee implants that promote better osseointegration and have a longer lifespan. The architecture and construction industries are also adopting these tools to design lighter, stronger, and more material-efficient structures, from complex building facades to long-span bridges, blending architectural beauty with engineering efficiency.

7.3 FABRICATION TECHNIQUES

7.3.1 Lithography

Lithography is one of the oldest and most widely used techniques for patterning micro- and nanoscale structures. The method relies on transferring a pre-designed pattern onto a substrate using light, electrons, or other radiation sources. Photolithography, the most common variant, uses ultraviolet (UV) light to expose a photoresist layer coated on the substrate. The exposed regions undergo chemical changes, allowing selective removal and subsequent material deposition or etching. Electron beam lithography (EBL) extends this principle to achieve higher resolution, reaching sub-10-nanometer features. Lithography is particularly effective for producing periodic and highly intricate patterns, such as those found in split-ring resonators, wire arrays, and photonic crystals.

One of the primary advantages of lithography is its precision and scalability. It allows the production of uniform structures over large areas while maintaining nanoscale resolution. This makes it ideal for high-frequency metamaterials, semiconductor devices, and integrated photonic circuits. However, the technique also has limitations. Photolithography is constrained by the wavelength of the light used, necessitating expensive deep-UV or extreme-UV systems for very small features. Electron beam lithography, while highly precise, is time-consuming and unsuitable for large-scale production. Additionally, lithography often requires multiple processing steps, including resist coating, exposure, development, etching, and cleaning, each of which introduces potential variability. Despite these challenges, lithography remains a foundational technique in modern micro- and nanoscale fabrication.

7.3.2 3D Printing

Three-dimensional printing, also known as additive manufacturing, has revolutionized the way complex structures are fabricated. Unlike traditional subtractive methods, 3D printing builds structures layer by layer based on computer-aided design (CAD) models. In the context of metamaterials, 3D printing allows the creation of intricate geometries, including lattice structures, hollow channels, and freeform shapes that would be challenging or impossible with conventional methods. Techniques such as stereolithography (SLA), selective laser sintering (SLS), and two-photon polymerization enable printing at micro- and nanoscale resolutions.

The primary advantage of 3D printing lies in its flexibility. Designers can iterate quickly, test multiple geometries, and optimize structures without the need for expensive masks or molds. Additionally, 3D printing supports multi-material fabrication, allowing combinations of metals, polymers, and composites within a single device. This capability is particularly important for multifunctional metamaterials, where mechanical, thermal, or optical properties must be integrated within a single structure. However, challenges remain in achieving the high resolution required for certain applications, particularly in optical or electromagnetic metamaterials operating at visible or infrared wavelengths. Post-processing steps, such as surface smoothing or sintering, may be necessary to achieve the desired material properties. Despite these challenges, 3D printing is increasingly recognized as a versatile tool for rapid prototyping and custom fabrication.

7.3.3 Nanoimprinting

Nanoimprinting is a high-resolution, cost-effective technique for replicating nanoscale patterns over large areas. The process involves pressing a hard mold or stamp containing nanoscale features into a soft polymer or resist layer on a substrate. After imprinting, the material is hardened, typically through thermal curing or UV exposure, and the mold is removed, leaving a precise negative replica of the pattern. Nanoimprinting is particularly suited for creating periodic structures such as photonic crystals, plasmonic arrays, and metamaterial surfaces.

A significant advantage of nanoimprinting is its ability to produce uniform nanoscale features over large areas with relatively low cost and high throughput. Unlike electron beam lithography, which is serial and time-intensive, nanoimprinting is inherently parallel, allowing rapid fabrication. This makes it suitable for commercial applications where scalability is crucial.

Additionally, nanoimprinting is compatible with flexible substrates, enabling the production of bendable or wearable devices. However, the technique also has limitations. Mold fabrication is complex and requires high precision, as defects or wear in the mold can propagate across all imprinted patterns. Alignment during imprinting is critical, particularly for multi-layered structures, and careful control of temperature, pressure, and material viscosity is necessary to ensure fidelity.

7.4 MATERIAL CHOICES AND CHALLENGES

7.4.1 Polymers

Polymers have emerged as one of the most versatile classes of materials in the design and fabrication of metamaterials. Their widespread availability, lightweight nature, and compatibility with modern fabrication techniques make them highly attractive for applications that demand flexibility, tunability, and cost-effectiveness. Unlike metals and ceramics, which often require high-temperature processing or exhibit brittleness, polymers offer a unique balance of structural adaptability and functional customization. In metamaterials, polymers serve not only as structural matrices but also as functional dielectrics, substrates, or hosts for reinforcing elements. However, the use of polymers is not without limitations, and understanding their challenges is crucial for advancing their role in next-generation metamaterial systems.

- **The Role of Polymers in Metamaterials:** Polymers are primarily valued for their dielectric properties, mechanical flexibility, and ease of processing. They can be engineered to exhibit specific refractive indices or dielectric constants, which are essential for controlling electromagnetic interactions in microwave and terahertz metamaterials. Their low density enables the creation of lightweight structures, an advantage in aerospace, defense, and wearable technologies. Furthermore, polymers are highly compatible with fabrication methods such as lithography, soft lithography, nanoimprinting, and 3D printing, allowing complex architectures to be realized with high precision and scalability. These capabilities make polymers an integral component of many practical metamaterial systems.
- **Electrical and Electromagnetic Properties:** From an electromagnetic perspective, polymers are typically insulators with low dielectric losses at microwave and terahertz frequencies. This makes them excellent candidates for use as substrates or embedding materials in split-ring resonators, wire arrays, or metasurfaces. By incorporating

conductive fillers such as carbon nanotubes, graphene flakes, or metallic nanoparticles, polymers can be transformed into tunable composites with enhanced electromagnetic responses. For example, polymer composites can be engineered to achieve specific permittivity values, enabling better impedance matching or resonance control in electromagnetic metamaterials. Despite this versatility, polymers face challenges at optical frequencies, where their dielectric behavior may limit performance compared to inorganic materials.

- **Mechanical Advantages and Flexibility:** One of the most significant advantages of polymers lies in their mechanical properties. They are inherently flexible, stretchable, and lightweight, which is particularly valuable in applications that require conformal or wearable devices. Metamaterials designed with polymer substrates can be bent, twisted, or stretched without significant degradation of their performance, enabling real-world integration into flexible electronics, biomedical sensors, and soft robotics. This adaptability is difficult to achieve with metals or ceramics, which are often rigid or brittle. However, the softness of polymers also means they are more prone to mechanical deformation under load, which can introduce stability issues in long-term use.
- **Thermal Behavior of Polymers:** The thermal properties of polymers present both advantages and challenges. Polymers generally have low thermal conductivity, which makes them effective thermal insulators in certain contexts but limits their ability to dissipate heat in high-power devices. Their relatively low glass transition and melting temperatures can also restrict their use in environments involving high thermal stress. For instance, polymer-based metamaterials may deform or degrade when exposed to prolonged heating, reducing their reliability in aerospace or industrial settings. Strategies to address these limitations include incorporating thermally conductive fillers such as boron nitride, graphene, or ceramic particles, which enhance heat dissipation while maintaining polymer flexibility.
- **Chemical Stability and Environmental Durability:** Another critical factor in the use of polymers is their chemical and environmental stability. While many polymers are chemically resistant to solvents, acids, and bases, others can degrade when exposed to ultraviolet radiation, oxygen, or moisture. This degradation often manifests as discoloration, embrittlement, or loss of mechanical integrity, ultimately reducing the performance of polymer-based metamaterials. Protective

coatings or chemical modifications are often required to extend their lifespan in outdoor or harsh environments. Furthermore, polymers are sensitive to creep and aging effects, which may affect the dimensional stability of metamaterials over time. Addressing these challenges requires careful selection of polymer types and surface treatments to ensure durability and reliability.

- **Fabrication Compatibility and Scalability:** Polymers are particularly well-suited for scalable and cost-effective fabrication. Techniques such as soft lithography, roll-to-roll processing, and additive manufacturing allow for large-area production of polymer-based metamaterials at relatively low cost. This scalability opens the door for commercial applications such as communication devices, sensors, and flexible displays. Nanoimprinting techniques further enable the creation of fine nanostructures on polymer surfaces, essential for photonic and plasmonic metamaterials. However, achieving nanoscale precision with polymers can be challenging due to issues like shrinkage, surface roughness, and thermal expansion. These fabrication-related challenges must be addressed to fully exploit the potential of polymers in high-frequency metamaterial devices.
- **Environmental and Sustainability Considerations:** The sustainability of polymer-based metamaterials is an increasingly important topic. Many commonly used polymers are derived from petrochemical sources and are not biodegradable, raising environmental concerns. As metamaterials move closer to commercialization, their environmental footprint becomes more relevant. Research into bio-based and biodegradable polymers is growing, with promising candidates such as polylactic acid (PLA) offering environmentally friendly alternatives. However, these sustainable polymers often have inferior mechanical or thermal properties compared to traditional polymers, requiring reinforcement strategies to make them viable for advanced metamaterial applications.
- **Challenges of Polymer Integration:** Despite their advantages, integrating polymers into metamaterial systems is not straightforward. Their relatively low mechanical strength can result in structural instability, particularly in micro- and nanoscale architectures. Thermal expansion mismatches between polymers and other materials, such as metals or ceramics, can lead to delamination or cracking under thermal cycling. Additionally, ensuring uniform dispersion of fillers in polymer composites is technically challenging, as agglomeration can lead to

localized defects and poor electromagnetic performance. Overcoming these integration challenges requires advances in polymer chemistry, composite processing, and interface engineering to achieve stable and reliable metamaterial designs.

7.4.2 Metals

Metals have played a fundamental role in the evolution of metamaterials, providing the necessary electromagnetic and structural properties that enable the exotic behaviors these engineered materials exhibit. Their excellent electrical conductivity, strong interaction with electromagnetic radiation, and mechanical robustness make metals indispensable in designing metamaterials across a wide range of frequencies. From microwave to optical regimes, metals serve as the backbone for structures such as split-ring resonators, wire arrays, and plasmonic devices. However, metals also present significant challenges that limit their performance and scalability.

- **The Role of Metals in Metamaterials:** Metals are crucial in metamaterials because they provide the negative permittivity required to manipulate electromagnetic waves in ways not achievable with natural materials. In structures such as split-ring resonators, metals act as conduits for induced currents, enabling resonance effects that produce negative permeability. Wire arrays made of metals achieve negative permittivity, a key condition for realizing left-handed materials and negative refractive index behavior. At optical frequencies, metals like gold and silver support surface plasmon resonances, which allow electromagnetic fields to be confined below the diffraction limit. These capabilities position metals as central elements in electromagnetic metamaterial design.
- **Electromagnetic Properties of Metals:** The electromagnetic response of metals arises from their free electron density, which allows them to reflect and absorb electromagnetic waves efficiently. At microwave frequencies, metals behave as near-perfect conductors with minimal energy loss, enabling efficient waveguiding and resonance. However, as frequencies approach the terahertz and optical ranges, metals begin to exhibit significant losses due to electron scattering, resistive heating, and interband transitions. This absorption reduces the efficiency of plasmonic metamaterials and limits their practical applications. Overcoming these losses remains one of the most significant challenges in metal-based metamaterial research. Alloying, nanostructuring, and hybridization with dielectrics are common strategies to mitigate these losses and improve performance.

- **Mechanical Strength and Structural Stability:** Beyond their electromagnetic function, metals contribute mechanical strength and stability to metamaterial structures. Metals such as copper, aluminum, and nickel are not only conductive but also durable, allowing metamaterials to maintain their structural integrity under stress. This robustness makes metals suitable for applications in aerospace, defense, and high-stress industrial environments where mechanical reliability is as important as electromagnetic performance. However, the density and weight of metals can pose limitations for lightweight and portable applications. For example, in wearable or aerospace devices, where minimizing mass is crucial, metallic metamaterials may require integration with lighter materials such as polymers or composites.
- **Thermal Properties of Metals:** Metals exhibit high thermal conductivity, a property that is advantageous in applications requiring efficient heat dissipation. In high-power electromagnetic devices, where resistive losses can generate significant heat, metallic components help prevent overheating and ensure reliable operation. However, the same property that makes metals good heat conductors also means that they are sensitive to thermal expansion. Mismatches in thermal expansion coefficients between metals and other materials can cause deformation, delamination, or cracks in hybrid metamaterial structures. Moreover, at very high operating temperatures, metals may soften or lose their mechanical integrity, limiting their use in extreme environments without proper reinforcement or alloying.
- **Chemical Stability and Environmental Challenges:** A major drawback of metals in metamaterials is their susceptibility to oxidation, corrosion, and chemical degradation. Copper and aluminum, for example, can oxidize in ambient conditions, leading to changes in surface conductivity and reduced electromagnetic performance. Gold and silver are more chemically stable, but their high cost limits large-scale or commercial use. Protective coatings, passivation layers, or encapsulation strategies are often necessary to ensure long-term stability of metal-based metamaterials in real-world environments. However, these additional steps can complicate fabrication and increase production costs. Ensuring chemical durability without sacrificing electromagnetic performance remains an ongoing challenge.
- **Fabrication Compatibility and Precision:** The use of metals in metamaterials is closely tied to advanced fabrication methods such as lithography, electroplating, and focused ion beam milling. Metals can

be patterned into nanoscale structures with high precision, which is essential for plasmonic metamaterials operating at optical frequencies. However, achieving uniformity and scalability in metallic nanostructures is technically challenging and often expensive. Furthermore, metals can be difficult to integrate with flexible substrates, limiting their use in stretchable or wearable metamaterials. Additive manufacturing and 3D printing of metals are emerging techniques that may overcome some of these limitations by enabling rapid prototyping and complex geometries. Yet, these methods still face challenges in achieving the fine resolution required for nanoscale metamaterial features.

- **Losses and Efficiency Limitations:** One of the most persistent challenges in using metals for metamaterials is the issue of ohmic and optical losses. At high frequencies, particularly in the visible and near-infrared spectrum, metals absorb a significant portion of incident electromagnetic energy. This results in heating and reduced efficiency of devices such as superlenses, cloaking structures, and plasmonic waveguides. Researchers are exploring alternative low-loss materials, including transparent conductive oxides and high-index dielectrics, to complement or replace metals in certain metamaterial applications. Hybrid designs that combine metals with dielectric or polymer structures are also being developed to balance conductivity with reduced absorption.
- **Cost and Scalability Issues:** The economic aspect of metal use in metamaterials cannot be ignored. While metals like copper and aluminum are relatively inexpensive, noble metals such as gold and silver, which are favored for their superior plasmonic properties, are costly and scarce. Large-scale production of metamaterials using such metals is not economically viable for most commercial applications. Developing low-cost alternatives or minimizing the amount of expensive metal required through nanostructuring or thin-film deposition is a critical research direction. Scalability also remains a challenge, as many fabrication techniques for metal-based metamaterials are suited for laboratory-scale prototypes rather than industrial-scale manufacturing.

7.4.3 Ceramics

Ceramics have long been valued for their unique combination of mechanical strength, thermal stability, and chemical resistance. In the context of metamaterials, ceramics are increasingly recognized as essential materials for applications where metals or polymers fall short. Their excellent dielectric properties, high-temperature tolerance, and resistance to harsh environments make them highly suitable for electromagnetic, acoustic, and thermal metamaterials. While ceramics provide performance advantages that few other material classes can match, they also present challenges such as brittleness, processing complexity, and difficulties in integration with other materials.

- **The Role of Ceramics in Metamaterials:** Ceramics serve multiple roles in metamaterials depending on the application. Their high dielectric constants make them useful in electromagnetic metamaterials operating at microwave, terahertz, and optical frequencies. They can function as resonators, substrates, or even active components in photonic crystals and dielectric metamaterials. In thermal metamaterials, ceramics are prized for their ability to withstand high operating temperatures without degradation, making them indispensable in cloaking and heat management systems. Their mechanical hardness also makes them valuable in vibration control and acoustic metamaterials, where durability and reliability are critical. These diverse applications underscore the versatility of ceramics as functional materials in metamaterial design.
- **Electromagnetic Properties of Ceramics:** The dielectric properties of ceramics are among their most valuable attributes in metamaterials. Many ceramics exhibit low dielectric loss and stable permittivity, enabling efficient manipulation of electromagnetic waves. Materials such as barium titanate, strontium titanate, and alumina are widely used in microwave and terahertz metamaterials for their ability to concentrate and guide electromagnetic fields. Unlike metals, ceramics do not suffer from significant ohmic losses, which makes them highly advantageous at higher frequencies. This allows ceramic-based metamaterials to operate with greater efficiency in photonic and optical applications. However, tailoring the dielectric properties of ceramics often requires careful compositional engineering, and their inherent brittleness can complicate the fabrication of delicate nanostructures.
- **Mechanical Strength and Brittleness:** Ceramics are well-known for their exceptional hardness and compressive strength, which make them attractive for durable metamaterial structures. Their mechanical robustness ensures long-term stability even under stress or harsh operating conditions. This is particularly important in acoustic metamaterials, where ceramics are used to construct resonant cavities or phononic crystals capable of withstanding repeated mechanical vibrations. However, ceramics are inherently brittle, meaning they can fracture easily under tensile stress or impact. This brittleness

limits their use in flexible or dynamic metamaterials, where polymers or composites may offer more reliable mechanical performance. Strategies such as creating ceramic composites or integrating ceramics with ductile materials are often employed to mitigate this limitation.

- **Thermal Properties of Ceramics:** Ceramics are renowned for their excellent thermal properties. Many ceramics maintain structural integrity at extremely high temperatures, often exceeding those tolerated by metals or polymers. This thermal stability makes ceramics ideal for high-temperature metamaterial applications, such as aerospace heat shields, energy systems, and thermal cloaking devices. Ceramics can also be engineered to have high or low thermal conductivity depending on the application. For example, materials like silicon carbide provide excellent heat conduction, while others such as zirconia act as thermal insulators. Despite these advantages, ceramics are susceptible to thermal shock, which can cause cracking when subjected to rapid temperature changes. Careful design and material selection are required to balance conductivity, stability, and resistance to thermal stress.
- **Chemical Stability and Environmental Resistance:** One of the most significant advantages of ceramics is their chemical inertness. Ceramics resist oxidation, corrosion, and degradation even in harsh chemical environments. This property allows ceramic-based metamaterials to operate in settings where metals or polymers would fail, such as in chemical processing plants, outer space, or environments with extreme humidity or radiation. Their stability ensures long-term reliability, making them ideal for applications that demand durability. However, ceramics can be difficult to bond with other materials due to their chemically inert surfaces, which limits their integration into hybrid metamaterial systems without additional processing steps such as surface treatments or specialized adhesives.
- **Fabrication Challenges and Compatibility:** The fabrication of ceramic metamaterials poses significant challenges. Traditional ceramic processing methods, such as sintering or hot pressing, often involve high temperatures and result in limited control over fine structural details. While these methods are effective for bulk materials, they are less suited for the nanoscale precision required in photonic and plasmonic metamaterials. Advances in additive manufacturing, such as ceramic 3D printing and stereolithography, have begun to address this challenge by enabling the production of complex geometries with

higher precision. However, these technologies are still developing and can be costly or limited in resolution. Integrating ceramics with other materials in composite or hybrid metamaterials also requires overcoming mismatches in thermal expansion and mechanical properties, which can lead to stress and failure at interfaces.

- **Acoustic Applications of Ceramics:** Ceramics have found particular success in acoustic metamaterials and phononic crystals. Their high density and rigidity make them effective for manipulating sound waves and creating bandgaps that block specific frequencies. Ceramic-based acoustic metamaterials are used in noise reduction, vibration isolation, and ultrasonic imaging systems. Their durability ensures consistent performance over long periods, even under repeated mechanical stress. However, their brittleness limits their ability to withstand impacts or dynamic bending, which can reduce their usefulness in portable or wearable acoustic devices. Hybrid designs that incorporate polymers or composites alongside ceramics are often used to achieve both durability and flexibility.
- **Economic and Scalability Considerations:** The economic feasibility of ceramic-based metamaterials is another important factor. While raw ceramic materials are often abundant and relatively inexpensive, the high cost of processing and fabrication can make ceramic metamaterials less economically attractive than polymer- or metal-based alternatives. Achieving nanoscale precision in ceramic structures requires advanced equipment and specialized processes, which may not be easily scalable for mass production. This limitation restricts ceramics primarily to specialized or high-performance applications, such as aerospace, defense, or energy systems, where their unique properties justify the higher cost. Research into more cost-effective ceramic processing methods will be crucial for broader adoption of ceramic metamaterials.

7.4.4 Composites

Composites are engineered materials that combine two or more distinct constituents to achieve properties that cannot be realized by a single material alone. In the context of metamaterials, composites offer unparalleled versatility, as they can integrate the benefits of polymers, metals, ceramics, or even nanomaterials into one system. By carefully designing the microstructure and tailoring the composition, researchers can produce composites that exhibit superior strength-to-weight ratios, enhanced thermal stability, or unique electromagnetic responses. The ability to customize composites makes them one of the most promising material choices for advancing metamaterial technologies.

- **The Role of Composites in Metamaterials:** Composites play a central role in metamaterial design due to their adaptability. They can be engineered to meet specific demands in electromagnetic, thermal, mechanical, and acoustic domains. For instance, fiber-reinforced composites can provide the mechanical strength and stiffness needed for structural metamaterials, while polymer-matrix composites embedded with metallic nanoparticles can deliver tunable electromagnetic properties. By combining materials with contrasting properties, composites allow researchers to achieve functionalities such as negative refractive indices, cloaking effects, and programmable responses that are difficult to obtain with single-phase materials. This ability to merge multiple characteristics into one system is why composites are often considered the backbone of practical metamaterial applications.
- **Electromagnetic Performance of Composites:** One of the most prominent applications of composites in metamaterials lies in electromagnetic wave manipulation. Composites can incorporate metallic inclusions for conductivity and resonant behavior while using polymer or ceramic matrices to provide dielectric stability. This combination allows for the creation of materials with tailored permittivity and permeability, which are essential for achieving negative index behavior or frequency-selective properties. For example, composites that integrate metal nanowires or nanoparticles in a dielectric medium can support plasmonic resonances, enabling advanced optical metamaterials. Similarly, carbon-fiber-reinforced composites offer conductivity alongside structural strength, making them useful for multifunctional devices. The primary challenge in this domain is achieving uniform distribution of inclusions within the matrix, as agglomeration can lead to performance inconsistency and unwanted scattering losses.
- **Mechanical Advantages of Composites:** Composites are widely appreciated for their mechanical properties, particularly their high strength-to-weight ratios. This characteristic is particularly beneficial for metamaterials intended for aerospace, automotive, or portable devices where weight reduction without compromising strength is critical. Fiber-reinforced polymers, such as those incorporating carbon or glass fibers, can withstand mechanical loads while maintaining flexibility for structural metamaterials. Additionally, composites often exhibit excellent fatigue resistance, ensuring long-term stability in dynamic environments. Despite these advantages, composites can face challenges related to delamination, interfacial cracks, and stress concentration at the boundaries between the matrix and reinforcing

phases. Overcoming these issues requires precise control of processing conditions and the development of improved bonding techniques between phases.

- **Thermal Properties and Challenges:** The thermal behavior of composites is another area of interest in metamaterial applications. By combining constituents with different thermal conductivities, composites can be engineered to direct heat flow in specific patterns, a feature crucial for thermal cloaking and energy management metamaterials. For example, composites containing high-conductivity fillers like graphene or boron nitride can achieve anisotropic heat conduction, making them valuable for thermal management in electronics. At the same time, ceramic-filled composites can withstand high operating temperatures while maintaining structural integrity. The challenge, however, lies in maintaining stability under thermal cycling. Differences in thermal expansion between the matrix and filler can lead to internal stresses and eventual material degradation. Therefore, tailoring thermal compatibility among phases is essential for the long-term reliability of composite-based thermal metamaterials.
- **Acoustic and Vibration Applications:** In the field of acoustic metamaterials, composites provide opportunities to design structures that control sound and vibration in innovative ways. Lightweight composites embedded with resonant inclusions can create phononic crystals or sound-absorbing materials that block or redirect specific frequencies. Polymer-based composites can provide flexibility for wearable acoustic devices, while ceramic-reinforced composites add rigidity for industrial soundproofing applications. The ability to tailor density and elasticity through compositional choices makes composites particularly suitable for vibration isolation systems and noise reduction technologies. Nevertheless, ensuring consistent performance requires precise distribution of acoustic inclusions and control over damping characteristics, both of which can be challenging to achieve at scale.
- **Fabrication Flexibility and Challenges:** A significant advantage of composites is the flexibility of fabrication methods available to produce them. Techniques such as lamination, additive manufacturing, injection molding, and resin transfer molding allow for the creation of complex geometries and hierarchical structures that are essential in metamaterial design. This adaptability makes composites particularly suited for scaling up metamaterial production compared to ceramics or metals, which are often more challenging to shape. However, fabrication

challenges persist, especially when working at the nanoscale. Achieving precise alignment of fibers, nanoparticles, or inclusions within a matrix is technically demanding and often expensive. Moreover, defects such as voids, poor dispersion, or incomplete curing can compromise performance, underscoring the need for advanced processing technologies and quality control measures.

- **Interfacial Compatibility Issues:** One of the defining challenges of composites in metamaterials is interfacial compatibility between different phases. Strong adhesion between the matrix and reinforcement is critical for transferring stress, maintaining structural integrity, and achieving predictable functional performance. Poor bonding can lead to delamination, microcracks, and failure under mechanical, thermal, or electromagnetic stresses. Surface treatments, coupling agents, and nanostructured interfaces are often employed to improve interfacial bonding, but these add complexity to the fabrication process. Additionally, mismatches in properties such as thermal expansion, conductivity, or stiffness between phases can introduce localized stresses that reduce long-term stability. Developing strategies to address interfacial challenges is one of the most active areas of research in composite metamaterials.
- **Multifunctionality of Composites:** Perhaps the most compelling feature of composites is their ability to achieve multifunctionality. Unlike single-phase materials, composites can simultaneously address structural, thermal, electromagnetic, and acoustic requirements. For instance, a carbon-fiber-reinforced polymer composite can provide lightweight structural strength while incorporating conductive fillers to enable electromagnetic shielding or wave manipulation. Similarly, ceramic-polymer composites can combine thermal resistance with flexibility, offering new solutions for high-temperature wearable metamaterials. This multifunctionality positions composites as highly adaptable materials capable of bridging the gap between laboratory demonstrations and practical real-world applications. The challenge lies in balancing these functions without sacrificing one property for another, which often requires sophisticated design and optimization strategies.
- **Economic and Scalability Considerations:** From an economic perspective, composites offer both opportunities and limitations. On one hand, many composite systems are cost-effective due to the relative abundance of raw materials and the availability of scalable

manufacturing techniques. On the other hand, advanced composites with nanomaterial reinforcements, precise structural alignment, or multifunctional properties can be expensive to produce, particularly at industrial scales. Additionally, the recyclability of composites remains a pressing issue, as separating different phases at the end of life is often difficult. This environmental challenge is particularly relevant as sustainability becomes a key criterion in material selection for future technologies.

CHAPTER 8

Reconfigurable and Tunable Metamaterials

8.1 PHASE-CHANGE MATERIALS (PCMS)

A phase-change material (PCM) is a substance which releases/absorbs sufficient energy at phase transition to provide useful heat or cooling. Generally the transition will be from one of the first two fundamental states of matter - solid and liquid - to the other. The phase transition may also be between non-classical states of matter, such as the conformity of crystals, where the material goes from conforming to one crystalline structure to conforming to another, which may be a higher or lower energy state.

The energy required to change matter from a solid phase to a liquid phase is known as the enthalpy of fusion. The enthalpy of fusion does not contribute to a rise in temperature. As such, any heat energy added while the matter is undergoing a phase change will not produce a rise in temperature. The enthalpy of fusion is generally much larger than the specific heat capacity, meaning that a large amount of heat energy can be absorbed while the matter remains isothermic. Ice, for example, requires 333.55 J/g to melt, but water will rise one degree further with the addition of just 4.18 J/g. Water/ice is therefore a very effective phase change material and has been used to store winter cold to cool buildings in summer since at least the time of the Achaemenid Empire.

By melting and solidifying at the phase-change temperature (PCT), a PCM is capable of storing and releasing large amounts of energy compared to sensible heat storage. Heat is absorbed or released when the material changes from solid to liquid and vice versa or when the internal structure of the material changes; PCMs are accordingly referred to as latent heat storage (LHS) materials.

There are two principal classes of phase-change material: organic (carbon-containing) materials derived either from petroleum, from plants or from animals; and salt hydrates, which generally either use natural salts from the sea or from mineral deposits or are by-products of other processes. A third class is solid to solid phase change. PCMs are used in many different commercial applications where energy storage and/or stable temperatures are required, including, among others, heating pads, cooling for telephone switching boxes, and clothing.

By far the biggest potential market is for building heating and cooling. In this application, PCMs hold potential in light of the progressive reduction in the cost of renewable electricity, coupled with the intermittent nature of such electricity. This can result in a mismatch between peak demand and availability of supply. In North America, China, Japan, Australia, Southern Europe and other developed countries with hot summers, peak supply is at midday while peak demand is from around 17:00 to 20:00. This creates opportunities for thermal storage media.

There are two common ways that PCMs may be used: either passively, where the PCM is located so as to absorb and then release heat due to temperature difference, which is thereby moderated. In such applications, the PCM may be encapsulated, and integrated into the structure of the object or space that is to be conditioned. In some applications, especially when incorporation to textiles is required, phase change materials are micro-encapsulated. Micro-encapsulation allows the material to remain solid, in the form of small bubbles, when the PCM core has melted.

Alternatively, the PCM can be contained in a vessel, and heat flow to and from the PCM can be controlled by pumping a heat transfer fluid through a heat exchanger, generally immersed in the PCM within the vessel. In this case the system is a sub-category of "thermal battery" or "TES", thermal energy storage, which encompasses sensible heat storage as well.

8.1.1 Characteristics and Classification

Latent heat storage can be achieved through changes in the state of matter from liquid→solid, solid→liquid, solid→gas and liquid→gas. However, only solid→liquid and liquid→solid phase changes are practical for PCMs. Although liquid–gas transitions have a higher heat of transformation than solid–liquid transitions, liquid→gas phase changes are impractical for thermal storage because large volumes or high pressures are required to store the materials in their gas phase. Solid–solid phase changes are typically very slow and have a relatively low heat of transformation.

Initially, solid–liquid PCMs behave like sensible heat storage (SHS) materials; their temperature rises as they absorb heat. When PCMs reach their phase change temperature (their melting point) they absorb large amounts of heat at an almost constant temperature until all the material is melted.

When the ambient temperature around a liquid material falls, the PCM solidifies, releasing its stored latent heat. A large number of PCMs are available in any required temperature range from -5 up to 190 °C. Within the human comfort range between 20 and 30 °C, some PCMs are very effective, storing over 200 kJ/kg of latent heat, as against a specific heat capacity of around one kJ/(kg·°C) for masonry. The storage density can therefore be 20 times greater than masonry per kg if a temperature swing of 10 °C is allowed. However, since the mass of the masonry is far higher than that of PCM this specific (per mass) heat capacity is somewhat offset. A masonry wall might have a mass of 200 kg/m^2, so to double the heat capacity one would require additional 10 kg/m^2 of PCM.

Example of Organic Bio-Based PCM in a Poly/Foil Encapsulation for Durability in Building Applications, where it Works to Reduce HVAC Energy Consumption and Increase Occupant Comfort

Organic PCMs

Hydrocarbons, primarily paraffins (C_nH_{2n+2}) and lipids but also sugar alcohols.

- Advantages:
  - Freeze without much supercooling.
  - Ability to melt congruently.
  - Self-nucleating properties.
  - No segregation.
  - Chemically stable.
  - Safe and non-reactive.
- Disadvantages:
  - Low thermal conductivity in their solid state. High heat transfer rates are required during the freezing cycle. Nano composites were found to yield an effective thermal conductivity increase up to 216%.

- High cost of purification, e.g. through distillation, means that organics may not have a sharp freezing/melting point.
- Volumetric latent heat storage capacity can be low.
- Flammable. This can be partially alleviated by specialized containment.

Inorganic

Salt hydrates (MxNy·nH2O)

- Advantages:
  - High volumetric latent heat storage capacity.
  - Availability and low cost.
  - Sharp melting point.
  - High thermal conductivity.
  - High heat of fusion.
  - Non-flammable.
  - Sustainability.

Example: Eutectic Salt Hydrate PCM with Nucleation and Gelling Agents for Long-Term Thermal Stability and Thermoplastic Foil Macro-Encapsulation Physical Durability

- Disadvantages:
  - Difficult to prevent incongruent melting and phase separation upon cycling, which can cause a significant loss in latent heat enthalpy.

- Can be corrosive to many other materials, such as metals. This can be overcome by only using specific metal-PCM pairings or encapsulation in small quantities in non-reactive plastic.
- Change of volume is very high in some mixtures
- Super cooling can be a problem in solid–liquid transition, necessitating the use of nucleating agents which may become inoperative after repeated cycling

Hygroscopic Materials

Many natural building materials are hygroscopic, that is they can absorb (water condenses) and release water (water evaporates). The process is thus:

- Condensation (gas to liquid) $\Delta H<0$; enthalpy decreases (exothermic process) gives off heat.
- Vaporization (liquid to gas) $\Delta H>0$; enthalpy increases (endothermic process) absorbs heat (or cools).

Large surfaces area are required to transfer significant amounts of heat. This heat storage, especially day-time cooling, is one of the benefits of masonry construction and traditional finishes of stucco in Northern Europe and Northeast Asia, as well as renders of earth, clay and cow dung in South Asia.

Solid-Solid PCMs

A specialized group of PCMs that undergo a solid/solid phase transition with the associated absorption and release of large amounts of heat. These materials change their crystalline structure from one lattice configuration to another at a fixed and well-defined temperature, and the transformation can involve latent heats comparable to the most effective solid/liquid PCMs. Such materials are useful because, unlike solid/liquid PCMs, they do not require nucleation to prevent supercooling.

Additionally, because it is a solid/solid phase change, there is no visible change in the appearance of the PCM, and there are no problems associated with handling liquids, e.g. containment, potential leakage, etc. Currently the temperature range of solid-solid PCM solutions spans from -50 °C (-58 °F) up to +175 °C (347 °F). Therefore, these materials have emerged as promising alternatives to traditional solid/liquid PCMs due to their ability to undergo phase transitions without liquefaction. This property eliminates the risk of leakage and enhances material stability, For example, SSPCMs include polymer-based materials such as polyethylene glycol and metal-organic frameworks (MOFs). In addition, SSPCMs have been explored for use in smart textiles, electronics cooling systems, and thermally adaptive building materials. Research efforts continue to optimize their thermal storage density and improve long-term cycling stability, supporting broader commercial applications.

In particular, integrating SSPCMs with nanostructured material and composite frameworks is being investigated to enhance their thermal conductivity and phase transition kinetics.

8.1.2 Selection Criteria

The phase change material should possess the following thermodynamic properties:

- Melting temperature in the desired operating temperature range.
- High latent heat of fusion per unit volume.
- High specific heat, high density, and high thermal conductivity.
- Small volume changes on phase transformation and small vapor pressure at operating temperatures to reduce the containment problem.
- Congruent melting.

Kinetic properties:

- High nucleation rate to avoid supercooling of the liquid phase.
- High rate of crystal growth, so that the system can meet demands of heat recovery from the storage system.

Chemical properties:

- Chemical stability.
- Complete reversible freeze/melt cycle.
- No degradation after a large number of freeze/melt cycle.
- Non-corrosiveness, non-toxic, non-flammable and non-explosive materials.

Economic properties:

- Low cost.
- Availability.

8.1.3 Thermophysical Properties

Key thermophysical properties of phase-change materials include: Melting point (T_m), Heat of fusion (ΔH_{fus}), Specific heat (c_p) (of solid and liquid phase), Density (ρ) (of solid and liquid phase) and thermal conductivity. The thermal properties of representative PCMs are shown below. Values such as volume change and volumetric heat capacity can be calculated there from. One major challenge is the inherently low thermal conductivity of many PCMs, which limits their heat transfer efficiency. To address this problem, high thermal conductivity additives such as carbon nanotube, graphene, and metallic nanoparticles have been introduced to enhance their performance. Another critical issue is supercooling, where the PCM remains in a liquid state below its freezing point. Solutions such as nucleating agents and encapsulation techniques have been developed to mitigate this effect. Additionally, volume expansion during phase transitions can impact material stability, necessitating advanced structural designs and containment strategies. Recent studies have also explored nano-enhanced PCMs and composite structures to further optimize thermal response times and cycling stability.

Table: Nano-Enhanced PCMs, Particularly those Incorporating Metal Foams

PCM type	Material	Melting point (°C)	Latent heat (kJ/kg)	Thermal conductivity (W/m· K)
Organic PCMs	Paraffin wax	20-70	150-250	0.2-0.4
	Fatty acids	20-70	150-200	0.2-0.3
Inorganic PCMs	Hydrated salts	30-120	200-350	0.5-1.2
	Metallic PCMs	30-1000	100-300	10-50
	Eutectic salts	50-250	100-250	0.5-2.0
Solid-solid PCMs	Graphene-enhanced PCMs	30-150	150-350	1.0-10.0

8.1.4 Technology, Development and Encapsulation

The most commonly used PCMs are salt hydrates, fatty acids and esters, and various paraffins (such as octadecane). Recently also ionic liquids were investigated as novel PCMs.

As most of the organic solutions are water-free, they can be exposed to air, but all salt based PCM solutions must be encapsulated to prevent water evaporation or uptake. Both types offer certain advantages and disadvantages and if they are correctly applied some of the disadvantages becomes an advantage for certain applications. They have been used since the late 19th century as a medium for thermal storage applications. They have been used in such diverse applications as refrigerated transportation for rail and road applications and their physical properties are, therefore, well known.

Unlike the ice storage system, however, the PCM systems can be used with any conventional water chiller both for a new or alternatively retrofit application. The positive temperature phase change allows centrifugal and absorption chillers as well as the conventional reciprocating and screw chiller systems or even lower ambient conditions utilizing a cooling tower or dry cooler for charging the TES system. The temperature range offered by the PCM technology provides a new horizon for the building services and refrigeration engineers regarding medium and high temperature energy storage applications. The scope of this thermal energy application is wide-ranging of solar heating, hot water, heating rejection (i.e., cooling tower), and dry cooler circuitry thermal energy storage applications. Since PCMs transform between solid–liquid in thermal cycling, encapsulation naturally became the obvious storage choice.

- Encapsulation of PCMs:
  - Macro-Encapsulation: Early development of macro-encapsulation with large volume containment failed due to the poor thermal conductivity of most PCMs. PCMs tend to solidify at the edges of the containers preventing effective heat transfer.
  - Micro-Encapsulation: Micro-encapsulation on the other hand showed no such problem. It allows the PCMs to be incorporated into construction materials, such as concrete, easily and economically. Micro-encapsulated PCMs also provide a portable heat storage system. By coating a microscopic sized PCM with a protective coating, the particles can be suspended within a continuous phase such as water. This system can be considered a phase change slurry (PCS).
  - Molecular-Encapsulation is another technology, developed by Dupont de Nemours that allows a very high concentration of PCM within a polymer compound. It allows storage capacity up to 515 kJ/m^2 for a 5 mm board (103 MJ/m^3). Molecular-encapsulation allows drilling and cutting through the material without any PCM leakage.

As phase change materials perform best in small containers, therefore they are usually divided in cells. The cells are shallow to reduce static head – based on the principle of shallow container geometry. The packaging material should conduct heat well; and it should be durable enough to withstand frequent changes in the storage material's volume as phase changes occur. It should also restrict the passage of water through the walls, so the materials will not dry out (or water-out, if the material is hygroscopic). Packaging must also resist leakage and corrosion. Common packaging materials showing chemical compatibility with room temperature PCMs include stainless steel, polypropylene, and polyolefin. Nanoparticles such as carbon nanotubes, graphite, graphene, metal and metal oxide can be dispersed in PCM. It is worth noting that inclusion of nanoparticles will not only alter thermal conductivity characteristic of PCM but also other characteristics as well, including latent heat capacity, sub-cooling, phase change temperature and its duration, density and viscosity. The new group of PCMs called NePCM. NePCMs can be added to metal foams to build even higher thermal conductive combination.

8.1.5 Thermal Composites

Thermal composites is a term given to combinations of phase change materials (PCMs) and other (usually solid) structures. A simple example is a copper mesh immersed in paraffin wax. The copper mesh within paraffin wax can be considered a composite material, dubbed a thermal composite. Such hybrid materials are created to achieve specific overall or bulk properties (an example being the encapsulation of paraffin into distinct silicon dioxide nanospheres for increased surface area-to-volume ratio and, thus, higher heat transfer speeds).

Thermal conductivity is a common property targeted for maximization by creating thermal composites. In this case, the basic idea is to increase thermal conductivity by adding a highly conducting solid (such as the copper mesh or graphite) into the relatively low-conducting PCM, thus increasing overall or bulk (thermal) conductivity.

If the PCM is required to flow, the solid must be porous, such as a mesh. Solid composites such as fiberglass or kevlar prepreg for the aerospace industry usually refer to a fiber (the kevlar or the glass) and a matrix (the glue, which solidifies to hold fibers and provide compressive strength). A thermal composite is not so clearly defined but could similarly refer to a matrix (solid) and the PCM, which is of course usually liquid and/or solid depending on conditions. They are also meant to discover minor elements in the earth.

8.1.6 Photo-Thermal Conversion Phase-Change Composite Energy Storage Materials (PTCPCESMs)

PTCPCESMs are composite phase change materials with photo-thermal materials. They have wide applications in various industries, owing to their high thermal conductivity, photo-thermal conversion efficiency, latent heat storage capacity, physicochemical stability, and energy saving effect.

PTCPCESMs mainly consist of functional carrier materials and organic PCMs. During the solid-liquid phase transition, organic PCMs can absorb and release a large amount of latent heat. Meanwhile, functional carrier materials not only enhance the stability and efficiency of photo-thermal conversion but also introduce various energy conversion functions. The photo-thermal conversion is related to the band structure and other electric properties of photo-thermal materials, contributing to different absorbing solar spectrum. This is achieved using materials like carbon-based nanostructures (e.g., graphene, CNTs), plasmonic nanoparticles (e.g., Au, Ag), and semiconductors (e.g., TiO_2, MoS_2). Common PCMs include organic materials (paraffins, fatty acids) and inorganic materials (salt hydrates, metal alloys).

Researchers have been working on high-efficiency PTCPCESMs. A combined form of difunctional phase change composites integrated with phase change materials and photothermal conversion materials can reach 51.25% photothermal conversion efficiency and show no leakage under 60 °C for 24 h. Some researchers synthesized a novel form-stable solar-thermal conversion and storage materials by incorporating amino-functionalized single-walled carbon nanotubes into a polyethyleneglycol based polyurethane PCM, and reached a solar thermal conversion and storage efficiency of 89.3%.

8.1.7 Recent Advances in Phase-Change Materials

- **High-Performance PCM Development:** Recent research has focused on enhancing the efficiency and stability of PCMs through material innovations. New organic-inorganic composite PCMs, such as paraffin-based microencapsulated systems and salt hydrates with enhanced thermal conductivity, have demonstrated improved energy storage capabilities. In addition, metal-organic frameworks(MOFs) has investigated as a potential PCM candidates due to their tunable phase transition properties and high thermal storage density.
- **Applications in Energy Storage and Management:** PCMs have been increasingly utilized in energy storage systems, particularly in renewable energy applications. One promising approach is the integrations of PCMs into thermal energy storage units for solar and wind power systems. By mitigating fluctuations in power generation, these materials enhance reliability of renewable energy sources. Furthermore, the incorporations of PCMs into lithium-ion battery systems has shown potential in managing thermal runaway, thereby improving battery safety and longevity. Additionally PCM-enhanced smart windows and walls have been developed to regulate indoor temperatures and reduce building energy consumption by up to 30%. PCM-integrated heat pump systems have also demonstrated significant savings in heating and cooling applications.

8.1.8 Applications

Applications of phase change materials include, but are not limited to:

- Thermal energy storage.
- Solar cooking.
- Cold-energy battery.
- Conditioning of buildings, such as 'ice-storage'.
- Cooling of heat and electrical engines.
- Cooling: food, beverages, coffee, wine, milk products, green houses.
- Delaying ice and frost formation on surfaces.
- Medical applications: transportation of blood, operating tables, hot-cold therapies, treatment of birth asphyxia.
- Human body cooling under bulky clothing or costumes.
- Waste heat recovery.
- Off-peak power utilization: Heating hot water and Cooling.

- Heat pump systems.
- Passive storage in bioclimatic building/architecture (HDPE, paraffin).
- Smoothing exothermic temperature peaks in chemical reactions.
- Solar power plants.
- Spacecraft thermal systems.
- Thermal comfort in vehicles.
- Thermal protection of electronic devices.
- Thermal protection of food: transport, hotel trade, ice-cream, etc.
- Textiles used in clothing.
- Computer cooling.
- Turbine Inlet Chilling with thermal energy storage.
- Telecom shelters in tropical regions. They protect the high-value equipment in the shelter by keeping the indoor air temperature below the maximum permissible by absorbing heat generated by power-hungry equipment such as a Base Station Subsystem. In case of a power failure to conventional cooling systems, PCMs minimize use of diesel generators, and this can translate into enormous savings across thousands of telecom sites in tropics.

8.1.9 Fire and Safety Issues

Salt hydrates phase change materials contain on the order of 50% water and are nonflammable. Generally comprising common salts, which are found in seawater, they are usually nontoxic (albeit corrosive), and can, in principle, be disposed of, if no longer wanted, by dumping in the sea, whence they came. Organic PCMs are generally flammable. Some are also toxic, and may contain small amounts of aromatics or other toxic fractions of petroleum refining, as distillation is not usually complete. As such, PCMs must be selected and applied very carefully, in accordance with fire and building codes and sound engineering practices. Because of the increased fire risk, smoke, potential for explosion when held in containers, it may be wise not to use flammable PCMs within residential or other regularly occupied buildings. A Safety Data Sheet should be requested from the supplier.

8.2 MEMS/NEMS-BASED TUNABLE SYSTEMS

8.2.1 Evolution of MEMS and NEMS in Metamaterial Research

The early stages of metamaterial development were focused largely on static structures that offered unusual properties such as negative refractive index or cloaking. While groundbreaking, these designs suffered from limited flexibility, as their responses were confined to narrow frequency bands or specific external conditions. The emergence of MEMS technology in the late twentieth century provided the first practical framework for introducing tunability into metamaterials. MEMS-based actuators, switches, and resonators were soon adapted to reconfigure the geometry of metamaterial unit cells, altering their electromagnetic resonance frequencies in real time. With the progression of nanotechnology, NEMS extended this principle to even smaller scales, allowing manipulation of optical and quantum phenomena. Together, MEMS and NEMS are now integral to the next generation of adaptive metamaterials.

8.2.2 Principles of MEMS/NEMS-Based Tunability

The tunability offered by MEMS and NEMS arises from their capacity to physically alter the geometry, spacing, or orientation of metamaterial elements with high precision. In MEMS-based systems, electrostatic, thermal, magnetic, or piezoelectric actuation mechanisms can be employed to reconfigure resonant structures such as split-ring resonators or wire arrays. For instance, by slightly changing the gap of a split-ring resonator using a MEMS actuator, the resonance frequency can be shifted across a wide spectral range. NEMS-based systems operate on similar principles but extend the scale down to nanometers, enabling control at optical and infrared frequencies where traditional MEMS would be too large or slow. This mechanical reconfigurability translates directly into tunable electromagnetic properties, granting unprecedented flexibility in designing adaptive materials.

8.2.3 Electromagnetic Applications of MEMS/NEMS Systems

Electromagnetic metamaterials have greatly benefited from MEMS and NEMS integration. MEMS-based switches embedded in metamaterial arrays allow dynamic reconfiguration of their effective permittivity and permeability, making it possible to create frequency-agile devices. For instance, MEMS-actuated metamaterial absorbers can selectively tune absorption bands, enabling adaptive stealth technologies or reconfigurable filters. In optical domains, NEMS-based resonators can modulate plasmonic responses, allowing devices to function as nanoscale modulators or switches for integrated photonic circuits. Such capabilities are particularly important in telecommunications, where tunable filters and phase shifters are critical for managing bandwidth and improving signal quality. The marriage of MEMS and NEMS with electromagnetic metamaterials therefore represents a key advancement in realizing practical, reconfigurable devices for a variety of applications.

8.2.4 Acoustic and Mechanical Applications

MEMS and NEMS technologies are not limited to electromagnetic control but extend effectively into acoustic and mechanical metamaterials as well. MEMS-based microstructures can be designed to alter acoustic resonances, creating tunable sound filters or vibration isolation platforms. NEMS systems, owing to their nanoscale precision, can engineer phonon pathways to manipulate sound at very small wavelengths, paving the way for advanced acoustic cloaking devices. These tunable systems are highly valuable for applications in sonar, architectural acoustics, and noise management. On the mechanical side, MEMS/NEMS-based actuators can dynamically adjust stiffness, density, or elastic properties of a metamaterial structure, allowing for adaptive vibration damping or programmable structural behavior. This convergence of mechanics and electronics opens opportunities for real-time control of vibrations in sensitive environments such as aerospace or medical devices.

8.2.5 Thermal Tunability and Energy Applications

Another emerging area of research is the use of MEMS/NEMS-based systems in thermal metamaterials. MEMS devices can reconfigure pathways for heat flow, allowing thermal cloaks and heat redistribution systems to adapt dynamically to external conditions. Similarly, NEMS devices can manipulate phonon transport at the nanoscale, making it possible to control thermal conductivity in unprecedented ways. These tunable thermal metamaterials have direct applications in thermal management of electronic devices, energy harvesting, and even stealth technology where thermal signatures must be concealed. Coupling MEMS/NEMS devices with thermoelectric or photothermal components further enhances their potential by enabling active energy harvesting systems that adapt to environmental variations.

8.2.6 Fabrication and Integration Challenges

Despite their remarkable potential, MEMS and NEMS-based tunable systems face several challenges, particularly in fabrication and integration. MEMS fabrication, which relies heavily on established semiconductor processes, allows for relatively scalable production but requires precise control to ensure uniform performance across arrays of devices. NEMS fabrication, being even smaller in scale, is considerably more demanding, requiring advanced lithography and nanofabrication techniques that remain costly and time-intensive. Integration of MEMS and NEMS with metamaterials also introduces compatibility issues, as the mechanical movement of actuators must align precisely with the desired electromagnetic or acoustic response. Achieving this level of integration without compromising reliability or durability is one of the central engineering hurdles in advancing tunable metamaterials.

8.2.7 Reliability and Stability Considerations

Reliability is a critical concern for MEMS/NEMS-based systems, particularly because their performance depends on repeated mechanical actuation. Wear and tear over many cycles can lead to mechanical fatigue, stiction, or electrical failure. Additionally, environmental factors such as temperature fluctuations, humidity, and contamination can degrade device performance. NEMS devices are particularly vulnerable to quantum-level effects and thermal noise, which can destabilize their operation. Addressing these issues requires innovations in materials, such as the use of robust coatings or self-healing polymers, as well as optimized designs that minimize mechanical stress. Enhancing the longevity and stability of MEMS/NEMS-based tunable systems will be essential for their deployment in real-world metamaterial applications.

8.2.8 Multifunctional Capabilities

An attractive feature of MEMS and NEMS in metamaterial design is their multifunctional capability. A single MEMS device can serve as both a structural component and an active actuator, while NEMS devices can integrate sensing, actuation, and signal processing into one nanoscale unit. This multifunctionality reduces system complexity and enhances efficiency. For example, MEMS resonators embedded in a metamaterial array can both tune its electromagnetic properties and sense environmental changes, creating a feedback loop for self-adaptive materials. In photonic systems, NEMS-based devices can act as both modulators and detectors, paving the way for compact and integrated optical platforms. The multifunctional potential of MEMS/NEMS systems underscores their role in pushing metamaterials toward practical, adaptive technologies.

8.3 GRAPHENE AND 2D MATERIAL INTEGRATION

8.3.1 Graphene as a Building Block for Metamaterials

Graphene's unique properties make it an ideal building block for metamaterial design. Its high carrier mobility allows for ultrafast electronic responses, while its ability to support surface plasmon polaritons in the terahertz and infrared ranges enables strong light-matter interactions at subwavelength scales.

Unlike metals traditionally used in plasmonics, graphene offers tunability through electrostatic gating or chemical doping, allowing its conductivity and optical response to be adjusted in real time. This makes graphene-based metamaterials inherently reconfigurable, in contrast to conventional static designs. Furthermore, graphene's atomically thin nature ensures compatibility with layered or stacked metamaterial architectures, reducing weight and facilitating nanoscale integration without compromising functionality.

8.3.2 Integration of Graphene with Electromagnetic Metamaterials

One of the most prominent areas of research is the integration of graphene with electromagnetic metamaterials. Graphene sheets incorporated into split-ring resonators, metasurfaces, or photonic crystals enable active tuning of resonance frequencies and absorption characteristics. For example, by applying an external voltage, the Fermi level of graphene can be shifted, altering its interaction with incident electromagnetic waves. This feature has been exploited to design tunable absorbers, modulators, and sensors operating in the terahertz and mid-infrared regimes. Graphene also enhances light confinement and guides surface plasmons more efficiently than noble metals, making it invaluable for nanoscale photonic circuits. The ability to actively control electromagnetic responses places graphene-based metamaterials at the forefront of reconfigurable photonics and optoelectronics.

8.3.3 Graphene in Terahertz and Infrared Applications

Graphene's integration into terahertz and infrared devices addresses one of the longstanding challenges in metamaterials: the lack of efficient tunable materials in these frequency ranges. Conventional materials often exhibit weak interactions with terahertz radiation, limiting device performance. Graphene, however, supports tunable plasmonic modes in the terahertz band, making it possible to design modulators, detectors, and imaging systems that operate effectively in this spectrum. Similarly, in the mid-infrared range, graphene-enhanced metamaterials are used to build ultrasensitive detectors for chemical sensing and environmental monitoring, as many molecules have vibrational fingerprints in this spectral window. These applications demonstrate how graphene integration bridges the gap between theoretical metamaterial concepts and practical device performance.

8.3.4 2D Materials beyond Graphene

While graphene has dominated research in 2D materials, other members of this family bring complementary functionalities that expand the capabilities of metamaterials. Transition metal dichalcogenides such as MoS_2 and WS_2 possess direct bandgaps in the visible range, enabling strong excitonic effects and enhanced light absorption, which are crucial for optoelectronic devices. Hexagonal boron nitride serves as an excellent dielectric substrate for graphene, providing atomically smooth surfaces and acting as a hyperbolic material for mid-infrared photonics. Black phosphorus, with its anisotropic structure and tunable bandgap, offers opportunities for polarization-sensitive devices and photodetectors across a wide spectral range. Integrating these 2D materials with metamaterial architectures allows researchers to combine the tunability of graphene with bandgap engineering, nonlinear optical responses, and anisotropic control, creating multifunctional platforms with unprecedented versatility.

8.3.5 Metasurfaces and Flat Optics with 2D Materials

Metasurfaces, which are ultrathin planar structures that manipulate light at subwavelength scales, have emerged as a promising platform for integrating graphene and 2D materials. By embedding graphene layers within metasurfaces, researchers can achieve tunable reflection, transmission, and absorption properties. For instance, electrically gated graphene metasurfaces can function as dynamic beam steering devices, flat lenses, or holographic displays. Integration of TMDs within metasurfaces enhances nonlinear optical effects, enabling frequency conversion and all-optical switching. The combination of atomically thin 2D layers with flat optical platforms provides a pathway toward compact, lightweight, and energy-efficient devices for telecommunications, imaging, and quantum photonics.

8.3.6 Mechanical and Flexible Metamaterials with 2D Integration

Beyond optics and electromagnetics, 2D materials bring advantages to mechanical and flexible metamaterials. Graphene's exceptional strength and flexibility make it an ideal reinforcement material for mechanical metamaterials that require both structural integrity and lightweight performance. Additionally, 2D materials can be incorporated into flexible substrates to create stretchable, bendable metamaterials for wearable electronics and biomedical applications. The atomic thinness of graphene and TMDs ensures minimal impact on device thickness while significantly enhancing mechanical durability. This synergy is particularly valuable for applications where portability, flexibility, and robustness are essential, such as flexible sensors or adaptive acoustic systems.

8.3.7 Challenges in Integration of 2D Materials

Despite their enormous potential, the integration of graphene and other 2D materials into metamaterials faces substantial challenges. Large-scale production of defect-free 2D materials remains difficult, with methods such as chemical vapor deposition often resulting in grain boundaries or impurities that degrade performance. Ensuring uniform integration of 2D layers into complex metamaterial architectures is another significant obstacle, as even slight misalignments can alter device functionality. Additionally, maintaining the stability of 2D materials under environmental conditions is challenging, as materials like black phosphorus degrade rapidly in air. Addressing these challenges requires advances in fabrication techniques, encapsulation strategies, and transfer processes to ensure reliable and scalable integration.

8.3.8 Multifunctionality and Synergy of 2D Material Integration

One of the most exciting aspects of integrating 2D materials into metamaterials is the potential for multifunctionality. A single device can combine tunable electromagnetic responses, mechanical flexibility, and thermal control by strategically layering graphene, TMDs, and hBN. For instance, a composite metamaterial could exploit graphene for tunability, MoS_2 for excitonic effects, and hBN for dielectric support, resulting in a platform that is simultaneously reconfigurable, optically active, and structurally robust. This multifunctionality is particularly attractive for next-generation technologies such as 6G communication systems, quantum information processing, and multifunctional biomedical devices. The synergy achieved through 2D material integration sets the stage for a new paradigm in metamaterial design.

8.4 MAGNETIC AND ELECTRIC FIELD TUNING

8.4.1 Principles of Electric Field Tuning

Electric field tuning of metamaterials primarily exploits the electrostatic control of charge carriers within active materials integrated into metamaterial architectures. By applying a voltage bias, the carrier density in semiconductors, ferroelectrics, or two-dimensional materials such as graphene can be modulated, thereby altering their conductivity, permittivity, or refractive index.

This, in turn, changes the resonance condition of the metamaterial structure. Electric field control offers precise, reversible, and often ultrafast modulation, making it highly attractive for devices requiring dynamic switching. For instance, in terahertz metamaterials, electric gating of graphene layers embedded within resonator arrays can shift resonance frequencies and enhance absorption. Similarly, ferroelectric materials, with their spontaneous polarization states, allow strong tunability of dielectric properties under applied fields, enabling reconfigurable photonic crystals and metasurfaces.

8.4.2 Applications of Electric Field Tuning

The impact of electric field tuning is most evident in areas where rapid and broadband reconfigurability is essential. In communication systems, electrically tunable metamaterial antennas can dynamically adjust their radiation patterns, improving signal directionality and reducing interference. Electrically gated modulators and switches are crucial for photonic circuits operating in the infrared and optical domains, where integration with silicon technology is key for scalability. Sensing technologies also benefit, as the selective tuning of resonance frequencies enhances the detection of chemical and biological molecules with high sensitivity. Furthermore, energy harvesting systems can exploit electric field tuning to optimize resonance conditions for variable energy inputs, improving efficiency in dynamic environments.

8.4.3 Principles of Magnetic Field Tuning

Magnetic field tuning relies on the magneto-optical and magneto-dielectric effects present in certain materials. When exposed to an external magnetic field, materials such as ferrites, magnetic semiconductors, and multiferroics exhibit changes in permeability, refractive index, or anisotropy. These changes can be transferred to the metamaterial structure, modifying its resonance properties. Magnetic field control often provides strong tunability across wide frequency ranges, particularly in the microwave and terahertz bands. Unlike electric field tuning, which typically requires electrodes or direct contact, magnetic fields can influence metamaterial behavior non-invasively, which is advantageous for remote or harsh-environment applications.

8.4.4 Applications of Magnetic Field Tuning

Magnetic field tuning plays an important role in the development of nonreciprocal metamaterials, where electromagnetic wave propagation differs depending on direction. Such devices are key for isolators, circulators, and other components used in radar and communication systems. Magnetically tunable absorbers and filters allow for dynamic adaptation to varying signal environments, enhancing the performance of defense systems against jamming or stealth operations. Furthermore, in the terahertz regime, magnetoplasmonic metamaterials that integrate magnetic films with resonant structures enable active modulation of terahertz waves for spectroscopy and imaging. Magnetic tuning is also being investigated for spintronic and quantum devices, where the control of spin states through magnetic fields can be combined with metamaterial architectures to achieve novel functionalities.

8.4.5 Synergy of Magnetic and Electric Field Control

While electric and magnetic tuning mechanisms can be effective individually, their combination in multiferroic and hybrid metamaterials opens the door to greater multifunctionality. Multiferroic materials exhibit both ferroelectric and ferromagnetic order, enabling control of their properties through either electric or magnetic fields. By integrating such materials into metamaterial designs, devices can achieve dual-field tunability, providing redundancy, versatility, and enhanced control. For example, a hybrid metasurface incorporating graphene and ferrite layers could simultaneously exploit electric gating for ultrafast modulation and magnetic fields for broader spectral shifts. This synergy is particularly valuable for multifunctional devices where adaptability across multiple operating conditions is essential.

8.4.6 Advantages of Field-Tunable Metamaterials

Magnetic and electric field tuning offer several distinct advantages that elevate the performance and practicality of metamaterials. Electric field tuning provides precise and rapid control, with compatibility for integration into microelectronic and optoelectronic platforms. Magnetic field tuning, on the other hand, enables non-contact control, making it suitable for environments where electrodes are impractical. Together, these approaches make metamaterials adaptable to real-world conditions, where variability and unpredictability require systems capable of on-demand reconfiguration. Moreover, field-tunable metamaterials enable the reduction of device complexity by replacing multiple static components with a single adaptive element capable of performing multiple roles.

8.4.7 Challenges in Magnetic and Electric Field Tuning

Despite their promise, magnetic and electric field tuning approaches face significant challenges. For electric field tuning, the integration of active materials into metamaterial architectures often requires complex fabrication processes, and issues such as leakage currents, dielectric breakdown, and heating can limit performance. Additionally, maintaining uniform electric field distribution across large-scale metamaterials remains difficult. Magnetic field tuning, while powerful, generally requires strong external fields generated by bulky electromagnets, which can limit the portability and practicality of devices. Magnetic materials themselves may also suffer from losses at high frequencies, reducing the efficiency of tunable systems. Overcoming these challenges requires advances in material synthesis, nanofabrication, and compact field-generation technologies.

8.4.8 Emerging Trends in Field-Tunable Metamaterials

The future of magnetic and electric field tuning lies in the development of advanced materials and hybrid systems that enhance tunability while minimizing drawbacks. Two-dimensional materials such as graphene and transition metal dichalcogenides are already revolutionizing electric field control, and their integration with magnetic materials promises unprecedented multifunctionality.

Topological insulators and other quantum materials also offer unique opportunities for field-responsive metamaterials, with potential applications in quantum computing and secure communication. Additionally, miniaturized magnetic field sources, such as spintronic devices or nanoscale coils, are being explored to overcome the limitations of bulky electromagnets. The convergence of these trends points toward field-tunable metamaterials that are compact, efficient, and scalable for industrial applications.

8.5 SMART TEXTILES AND FLEXIBLE METAMATERIALS

8.5.1 Smart Textiles and Flexible Metamaterials

Smart textiles are fabrics engineered to sense, respond, and adapt to external stimuli, such as temperature, pressure, strain, or electromagnetic fields. When metamaterials are integrated into textiles, the resulting hybrid systems gain engineered wave-manipulation properties that can be dynamically tuned. Flexible metamaterials, on the other hand, are designed with mechanical compliance so that their functional performance is preserved even when subjected to bending, stretching, or twisting. These characteristics make them ideal for applications requiring comfort and adaptability, such as wearable devices, protective clothing, and implantable biomedical sensors.

8.5.2 Design Principles of Flexible Metamaterials

The design of flexible metamaterials focuses on incorporating structural motifs such as split-ring resonators, wire arrays, or metasurface patterns into soft, stretchable substrates. Instead of relying on rigid silicon wafers or ceramic platforms, flexible metamaterials use polymers, elastomers, and textile fibers as their base materials. This requires the careful design of patterns that maintain electromagnetic or mechanical responses under deformation. For example, serpentine or fractal patterns are often employed to ensure that resonant features remain stable during bending. The success of these designs depends on the seamless integration of structural functionality with mechanical durability, allowing the fabric or flexible surface to serve as both a wearable medium and a functional device.

8.5.3 Integration with Textiles

The integration of metamaterials into textiles involves techniques such as embroidery with conductive threads, printing conductive inks on fabric surfaces, or weaving functional fibers into the textile structure. These methods enable the embedding of resonant elements, antennas, or sensors directly into wearable fabrics. The advantage of this approach is that the functionality becomes inherently part of the fabric, eliminating the need for bulky attachments. For instance, a jacket with embedded metamaterial patches could dynamically regulate thermal conduction, enhance wireless communication, or provide electromagnetic shielding. Smart textiles that incorporate metamaterials therefore offer a seamless combination of everyday wearability and advanced performance.

8.5.4 Electromagnetic Applications

Flexible metamaterials and smart textiles have found substantial interest in the electromagnetic domain, particularly in communication and sensing. Wearable antennas made from textile-based metamaterials allow for improved signal strength and directionality while maintaining comfort and discretion. Metamaterial absorbers embedded in clothing can provide electromagnetic interference shielding, which is valuable in both civilian and military contexts. For defense applications, camouflage systems based on flexible metamaterials could adapt to different electromagnetic backgrounds, enhancing stealth. Furthermore, reconfigurable metasurfaces on flexible fabrics open the door to wireless power transfer, body-area communication networks, and enhanced connectivity in future 6G systems.

8.5.5 Mechanical and Acoustic Applications

Beyond electromagnetic control, flexible metamaterials incorporated into textiles can manipulate mechanical waves, vibrations, and sound. By embedding phononic structures into fabrics, wearable systems can achieve vibration damping, noise reduction, or even directional sound control. In industrial environments, such smart protective clothing could minimize worker exposure to harmful vibrations or acoustic noise. In sports and healthcare, garments with acoustic metamaterial properties can monitor joint or muscle activity by detecting and controlling sound or vibration signatures generated by body motion.

8.5.6 Thermal Regulation through Smart Textiles

One of the most promising applications of smart textiles with metamaterials is dynamic thermal regulation. Traditional fabrics passively insulate the body, but metamaterial-enhanced textiles can actively modulate heat flow. Thermal metamaterials embedded in flexible fibers can direct, concentrate, or dissipate heat depending on environmental conditions. This creates garments capable of cooling in hot environments and retaining warmth in cold conditions. Such adaptive thermal camouflage has significant military implications, as soldiers could become less detectable to infrared sensors. On the civilian side, energy-efficient clothing with built-in thermal management could reduce reliance on external heating or cooling systems.

8.5.7 Biomedical and Healthcare Applications

Smart textiles with flexible metamaterials are increasingly important in healthcare, where comfort, adaptability, and non-invasiveness are critical. Wearable sensors based on metamaterial structures can monitor physiological signals such as heart rate, respiration, blood oxygen levels, and body temperature. By embedding metamaterial resonators into clothing, real-time wireless sensing of these parameters becomes possible without bulky electronics.

Moreover, flexible metamaterials can enable non-invasive imaging, where electromagnetic signals interact with the body through fabric-integrated systems, aiding in early disease detection. Drug delivery systems and therapeutic devices are also being explored, where smart fabrics embedded with metamaterials can provide localized heating or electromagnetic stimulation for medical treatment.

8.5.8 Energy Harvesting in Wearable Systems

Another emerging direction is the integration of energy harvesting capabilities into smart textiles through metamaterials. Flexible metamaterial designs can be used to capture electromagnetic energy from ambient sources, such as Wi-Fi, radio waves, or even solar radiation. This harvested energy can then power wearable devices, reducing reliance on external batteries. Triboelectric and piezoelectric metamaterial designs incorporated into clothing can also convert mechanical motion into usable energy. This allows garments to become self-powered systems, capable of operating sensors, communication modules, or even small medical devices directly from the energy generated by the wearer's daily activities.

8.5.9 Challenges in Development

Despite their immense promise, smart textiles and flexible metamaterials face several challenges. Fabrication remains complex, as incorporating nanoscale resonant structures into large-area, stretchable fabrics requires scalable and cost-effective techniques. Durability is another critical issue, since clothing undergoes constant stress from bending, stretching, and washing. Ensuring that metamaterial performance remains stable under these conditions is an ongoing challenge. Additionally, the integration of power sources, data transmission systems, and user comfort must be carefully balanced to avoid bulky or rigid components. Biocompatibility and safety are also vital for wearable healthcare applications, as prolonged skin contact necessitates materials that are non-toxic and breathable.

CHAPTER 9

Emerging Applications

9.1 METAMATERIALS IN TELECOMMUNICATIONS

Telecommunications has always been at the forefront of technological innovation, driving progress in connectivity, information exchange, and global integration. From the invention of the telegraph and telephone to the development of fiber optics and wireless networks, advances in communication systems have transformed societies. In the twenty-first century, the demand for faster, more efficient, and more reliable communication continues to grow exponentially with the rise of mobile devices, the Internet of Things, satellite networks, and upcoming 6G technologies.

Against this backdrop, metamaterials have emerged as a transformative solution capable of redefining the way electromagnetic waves are manipulated, transmitted, and received in telecommunications. Their unique properties, derived not from their composition but from their structural design, enable functionalities that are difficult or impossible to achieve with conventional materials.

Metamaterials are artificial structures designed at the subwavelength scale to manipulate electromagnetic waves in unprecedented ways. Unlike natural materials, which have fixed permittivity and permeability values, metamaterials can be engineered to exhibit tailored refractive indices, including negative values.

This allows for the control of wave propagation, reflection, refraction, and absorption at multiple frequencies. In telecommunications, such control translates directly into improved antenna performance, reduced signal losses, enhanced data transmission speeds, and novel waveguiding capabilities. The ability to miniaturize devices while simultaneously expanding their performance spectrum makes metamaterials particularly valuable for communication systems operating from the microwave to the optical range.

- **Antenna Engineering with Metamaterials:** Antenna technology lies at the core of telecommunications, as antennas enable the transmission and reception of electromagnetic signals. Traditional antenna designs often face limitations related to size, efficiency, bandwidth, and directivity. Metamaterials provide solutions by introducing engineered substrates, superstrates, or metasurface-based antennas that improve performance while reducing footprint. By manipulating the effective refractive index, metamaterial-based antennas can achieve high directivity, beam steering, and enhanced gain without requiring large physical structures. This is particularly important for mobile devices, where space constraints demand compact designs. In addition, metamaterial-inspired antennas are reconfigurable, allowing them to dynamically adjust to different frequencies and communication standards, which is critical for future 6G networks and adaptive communication systems.

- **Waveguides and Transmission Lines:** Telecommunication systems rely heavily on waveguides and transmission lines to direct signals with minimal loss. Conventional waveguides often suffer from dispersion and leakage, which degrade signal quality. Metamaterials enable the design of left-handed transmission lines and subwavelength waveguides that can confine and guide electromagnetic waves with high efficiency. This allows for the miniaturization of communication circuits while maintaining or even enhancing performance. Plasmonic metamaterials at optical frequencies also provide means for nanoscale signal transmission, bridging the gap between electronic and photonic communication systems. Such innovations are paving the way for ultrafast on-chip interconnects and integrated photonic circuits, essential for data centers and high-capacity networks.
- **Electromagnetic Cloaking in Telecommunications:** One of the most intriguing applications of metamaterials is electromagnetic cloaking, which can also play a role in telecommunications. Cloaking devices are designed to guide electromagnetic waves around an object, rendering it invisible to detection. In communication systems, cloaking can be used to minimize signal interference by eliminating scattering from obstacles or by shielding devices from unwanted electromagnetic noise. This leads to cleaner signal transmission and improved efficiency, particularly in dense urban environments where interference from surrounding structures often poses challenges. While practical large-scale cloaking remains a challenge, partial cloaking and scattering reduction techniques based on metamaterials are already being applied in antenna systems.
- **Enhancing Bandwidth and Data Transmission:** As the demand for faster and more reliable communication grows, increasing bandwidth and data capacity becomes a critical objective. Metamaterials enable the development of broadband antennas and devices capable of operating across multiple frequency ranges. Their tunability also allows systems to adapt to different environments and user requirements. For example,

reconfigurable metasurfaces can dynamically adjust their resonance properties, providing variable bandwidths and enhancing spectral efficiency. This adaptability is crucial for the expansion of wireless communication technologies into higher frequency regimes such as millimeter waves and terahertz waves, which are considered essential for 6G systems.

- **Metasurfaces for Beam Steering and Holographic Communication:** Metasurfaces, a two-dimensional form of metamaterials, have gained considerable attention in telecommunications for their ability to manipulate electromagnetic wavefronts with ultrathin designs. By engineering subwavelength patterns, metasurfaces can achieve beam steering, polarization control, and wavefront shaping without relying on bulky mechanical systems. This makes them ideal for satellite communication, where lightweight and efficient beam steering is critical, and for 5G and 6G base stations, where fast and reconfigurable antennas are needed to support dynamic environments. Furthermore, metasurfaces open pathways toward holographic communication, where information is transmitted not only as data packets but also as structured waveforms that carry spatial and phase information, enhancing communication richness and security.
- **Terahertz Metamaterials in Next-Generation Networks:** Terahertz frequencies, which lie between microwave and infrared, have long been considered the frontier for communication technologies due to their enormous bandwidth potential. However, challenges such as high propagation losses and lack of efficient devices have hindered their adoption. Metamaterials are now addressing these barriers by enabling efficient terahertz sources, modulators, and detectors. Tunable terahertz metamaterials allow for dynamic control of wave propagation, opening new avenues for ultra-high-speed wireless communication. This is especially relevant for 6G and beyond, where data-intensive applications such as holographic telepresence, real-time virtual reality, and autonomous vehicle networks will require massive bandwidth and low-latency communication.

9.2 AEROSPACE AND DEFENSE SYSTEMS

The aerospace and defense sectors have always relied on technological superiority to ensure security, performance, and innovation. From stealth aircraft to high-speed communication satellites, advancements in materials science often play a decisive role in determining capabilities. Metamaterials, with their engineered properties that go beyond those found in nature, are rapidly emerging as a transformative class of materials in these industries. Their ability to manipulate electromagnetic waves, sound, mechanical vibrations, and thermal flows makes them uniquely suited for aerospace and defense applications, where control, adaptability, and performance are paramount.

9.2.1 Stealth and Electromagnetic Camouflage

One of the most prominent applications of metamaterials in defense is their use in stealth and electromagnetic camouflage. Traditional stealth technology relies on shaping aircraft and coating surfaces with radar-absorbing materials to reduce radar cross-sections. Metamaterials take this capability to the next level by enabling precise control over electromagnetic wave propagation.

With engineered refractive indices and subwavelength structures, metamaterial coatings can bend radar waves around an object, effectively cloaking it from detection. This goes beyond simple absorption and allows active manipulation of scattering signatures. Such advancements not only improve stealth aircraft but also enhance naval vessels, ground vehicles, and even soldier equipment, significantly increasing survivability in hostile environments.

9.2.2 Antennas and Communication Systems

Secure and reliable communication is central to defense and aerospace operations. Conventional antenna systems often face trade-offs between size, efficiency, and bandwidth. Metamaterial-based antennas overcome these limitations by providing compact, high-gain, and reconfigurable designs. These antennas can dynamically adapt to different frequencies and operational needs, which is critical for modern warfare where communication systems must handle rapidly changing conditions.

In aerospace, metamaterial antennas enable satellites and aircraft to communicate across multiple bands without bulky hardware. In defense, adaptive metamaterial-based antennas can enhance signal resilience against jamming and interception, ensuring robust communication in contested environments.

9.2.3 Sensors and Detection Systems

Metamaterials also hold significant promise for next-generation sensors used in aerospace and defense. Their ability to concentrate and manipulate electromagnetic fields at subwavelength scales leads to sensors with improved sensitivity and accuracy. For instance, terahertz metamaterial sensors can detect concealed weapons, explosives, or hazardous substances with high precision, providing enhanced security in airports and military checkpoints. In aerospace, metamaterial sensors can monitor structural integrity in aircraft or spacecraft, detecting minute cracks or stress points before they become catastrophic. This real-time monitoring contributes to improved safety and reduces maintenance costs.

9.2.4 Satellite and Space Applications

The aerospace industry increasingly relies on satellites for communication, navigation, Earth observation, and defense surveillance. Metamaterials play an essential role in improving the efficiency and functionality of satellite systems. Metamaterial-based reflectors and metasurfaces can be used to direct and focus signals more efficiently, reducing power requirements and enhancing signal strength. Lightweight metamaterial components also contribute to reducing payload mass, a critical factor in space missions. Additionally, metamaterials can improve the performance of solar panels through better light-trapping designs, enhancing energy harvesting for satellites operating in orbit. In the future, reconfigurable metamaterial devices may allow satellites to adapt their communication or sensing capabilities without hardware changes, extending mission lifetimes and versatility.

9.2.5 Acoustic and Vibration Control

In addition to electromagnetic applications, metamaterials also excel in controlling acoustic and mechanical vibrations, both of which are highly relevant in aerospace and defense. Acoustic metamaterials can be applied to reduce noise in aircraft cabins or to suppress sonar detection in submarines. Vibration-controlling metamaterials can be integrated into aircraft structures or spacecraft to dampen mechanical stresses, improving durability and stability during operations. For defense systems, vibration isolation through metamaterials is critical for precision equipment, such as missile guidance systems and sensitive electronics, where even minor disturbances can affect accuracy.

9.2.6 Thermal Management in Aerospace and Defense

Effective thermal management is vital in aerospace and defense applications, where systems often operate in extreme environments. Traditional thermal shielding materials have limitations in adaptability and efficiency. Thermal metamaterials, however, can control heat flow with unprecedented precision. By directing, concentrating, or dispersing thermal energy, they provide better insulation and heat dissipation for aircraft engines, spacecraft systems, or high-powered defense equipment. Thermal cloaking can even be applied to mask the infrared signatures of military vehicles, reducing their visibility to thermal imaging systems. These innovations not only enhance performance but also increase survivability in combat scenarios.

9.2.7 Radar and Imaging Systems

Metamaterials have also revolutionized radar and imaging technologies used in aerospace and defense. By exploiting subwavelength structures, metamaterial lenses and superlenses can overcome the diffraction limit, enabling near-field imaging with extraordinary resolution. This has implications for surveillance, reconnaissance, and target identification, where detecting small or distant objects can make a strategic difference. Additionally, metamaterials enhance the performance of synthetic aperture radars, which are widely used for mapping and monitoring. The ability to create tunable or reconfigurable radar systems also means that future defense platforms can adapt their detection capabilities in real time, making them more effective in complex operational environments.

9.2.8 Cybersecurity and Electromagnetic Shielding

With the increasing digitization of defense systems, protecting communication and information networks from cyber and electromagnetic threats has become critical. Metamaterials provide advanced shielding solutions by creating barriers that block or redirect unwanted electromagnetic interference. These can protect sensitive electronic systems from jamming, hacking, or surveillance. Smart metamaterial shields could be designed to allow only authorized signals to pass through, adding a new layer of security for military communication networks. In aerospace, shielding against cosmic radiation and electromagnetic interference is equally vital for maintaining spacecraft functionality and astronaut safety.

9.3 BIOMEDICAL IMAGING AND SENSING

Biomedical imaging and sensing are at the heart of modern healthcare and medical research. They allow clinicians and scientists to visualize, monitor, and analyze biological structures and processes in ways that were unimaginable a few decades ago. Techniques such as magnetic resonance imaging, ultrasound, optical microscopy, and biosensors have already revolutionized diagnostics and treatment. Yet, conventional methods often face limitations in resolution, penetration depth, sensitivity, and specificity. The emergence of metamaterials offers a new frontier for overcoming these challenges. By providing unprecedented control over electromagnetic, acoustic, and optical waves, metamaterials are driving innovations in biomedical imaging and sensing that promise earlier detection, more accurate diagnoses, and more effective therapies.

9.3.1 Promise of Metamaterials in Biomedicine

Metamaterials are artificially engineered structures with tailored properties not found in natural substances. Their unique ability to manipulate waves at scales smaller than their wavelength makes them highly attractive for biomedical applications. In imaging, metamaterials can overcome diffraction limits, enabling visualization of structures at the nanoscale. In sensing, they can enhance sensitivity by concentrating fields in extremely small regions, detecting subtle molecular or cellular changes. These capabilities are vital in medicine, where early detection of diseases like cancer or cardiovascular disorders can significantly improve patient outcomes.

9.3.2 Superlensing for Subwavelength Imaging

One of the most exciting contributions of metamaterials to biomedical imaging is the development of superlenses. Traditional optical systems are limited by diffraction, preventing visualization of objects smaller than half the wavelength of light. This restricts the resolution of biological imaging techniques such as light microscopy. Metamaterial-based superlenses can bend and refocus evanescent waves, which carry subwavelength details, thus capturing fine structures at the nanoscale. This capability is especially useful in cellular and molecular imaging, where researchers can observe proteins, organelles, or DNA interactions in real time. Superlensing has the potential to transform pathology, neuroscience, and microbiology by revealing details invisible to conventional techniques.

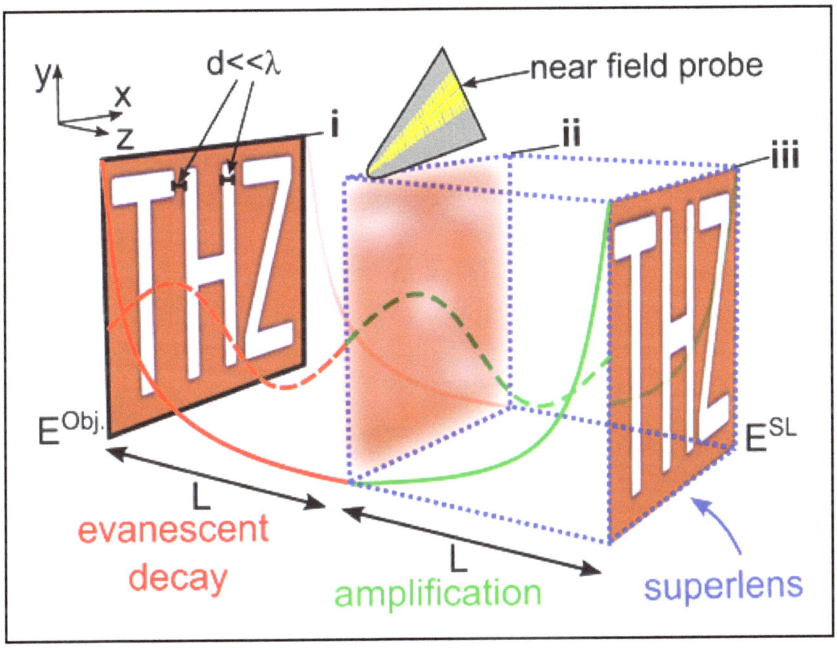

Subwavelength Terahertz Imaging via Virtual Superlensing in the Radiating Near Field

9.3.3 Plasmonic Metamaterials in Biomedical Sensing

Plasmonic metamaterials, which exploit the interaction of light with free electrons at metal surfaces, play a critical role in enhancing biomedical sensing. When light excites surface plasmons, it creates intense localized fields at the nanoscale. These fields are highly sensitive to changes in the surrounding environment, making plasmonic metamaterials powerful tools for detecting biomolecules. For example, plasmonic biosensors can detect trace amounts of biomarkers such as proteins or nucleic acids that indicate the presence of disease. Their extreme sensitivity allows for early detection of conditions such as cancer or viral infections, where conventional methods might miss low concentrations.

9.3.4 Terahertz Imaging and Metamaterials

Terahertz radiation lies between microwave and infrared frequencies and has long been considered promising for biomedical imaging. However, terahertz waves are difficult to generate and detect effectively due to material limitations. Metamaterials address this challenge by providing structures that can manipulate terahertz waves with high precision. In medical applications, terahertz metamaterial imaging can differentiate between healthy and diseased tissues based on their unique absorption spectra. For instance, cancerous tissues often exhibit distinct signatures in the terahertz range, allowing non-invasive and label-free detection. Moreover, terahertz sensing can identify hydration levels in tissues, offering valuable diagnostic insights into burns, skin diseases, and even dental health.

9.3.5 Acoustic and Ultrasound Metamaterials

Beyond electromagnetic waves, metamaterials also extend their utility to acoustic and ultrasound imaging, which are widely used in clinical settings. Ultrasound is valued for its safety and non-invasive nature, but it often struggles with resolution and depth limitations. Acoustic metamaterials can enhance ultrasound imaging by focusing sound waves more precisely or by creating negative refraction, leading to super-resolution imaging. In addition, metamaterial-based acoustic sensors can detect mechanical vibrations or pressure changes at the cellular level, opening new possibilities in cardiovascular monitoring and brain imaging. The combination of acoustic metamaterials with wearable devices could further enable continuous health monitoring outside of traditional clinical environments.

9.3.6 Magnetic Resonance Imaging Enhancement

Magnetic resonance imaging (MRI) is one of the most powerful diagnostic tools in medicine, providing high-resolution images of soft tissues. However, MRI is often constrained by low signal-to-noise ratios and long scanning times.

Metamaterials can enhance MRI performance by locally concentrating electromagnetic fields, thereby boosting signal strength and image clarity. Metamaterial-based resonators or lenses placed near the body can focus radiofrequency fields onto specific regions, improving resolution and reducing the time required for scans. This development is especially valuable in imaging small organs, early-stage tumors, or subtle neurological abnormalities that require high precision.

9.3.7 Non-Invasive Biosensing Applications

Metamaterial sensors are increasingly being explored for non-invasive biosensing applications that can revolutionize point-of-care diagnostics. By engineering structures that resonate at specific frequencies, metamaterials can detect changes in biological tissues, fluids, or even breath. For example, wearable metamaterial sensors integrated into smart textiles can continuously monitor physiological parameters such as glucose levels, hydration, or blood oxygen without the need for invasive sampling. Similarly, metamaterial-based breath analyzers can identify volatile organic compounds associated with diseases like lung cancer or diabetes. These advances align with the growing demand for personalized medicine and real-time health monitoring.

9.3.8 Integration with Nanotechnology and Biophotonics

The integration of metamaterials with nanotechnology and biophotonics further enhances their role in biomedical imaging and sensing. Nanostructured metamaterials can be functionalized with biomolecules to create hybrid devices that detect specific targets with high accuracy. In optical imaging, combining metamaterials with fluorescence techniques allows researchers to track cellular activity with unprecedented detail. Additionally, nanostructured metamaterials enable the creation of compact, lab-on-chip platforms that can perform rapid diagnostics using only tiny samples of blood or saliva. These miniaturized systems are particularly valuable for remote or resource-limited healthcare settings, where access to advanced laboratories is limited.

9.4 WIRELESS POWER TRANSFER AND ENERGY HARVESTING

Wireless power transfer and energy harvesting represent two of the most promising technological directions in modern energy systems. As devices become smaller, more mobile, and more interconnected through the Internet of Things, the demand for reliable, wireless, and sustainable sources of power grows rapidly.

Traditional battery technology, while widely used, faces limitations in capacity, recharge cycles, and environmental sustainability. Metamaterials, with their unique ability to manipulate electromagnetic waves in ways not possible with natural materials, have opened a new frontier in wireless power and energy harvesting. These engineered structures provide a pathway to more efficient energy transmission and collection, enabling breakthroughs in medical implants, wearable devices, telecommunications, and renewable energy systems.

9.4.1 Fundamentals of Wireless Power Transfer

Wireless power transfer, or WPT, is the process of transmitting electrical energy without direct physical connections. Traditional methods include inductive coupling, capacitive coupling, and resonant energy transfer, each of which relies on electromagnetic fields to move energy from a transmitter to a receiver. However, conventional systems are constrained by efficiency losses, limited transfer distance, and alignment sensitivity. Metamaterials address these challenges by focusing and guiding electromagnetic fields more effectively. By tailoring permittivity and permeability, metamaterials can enhance near-field coupling and extend the range of efficient power transfer. This capability is particularly significant in powering devices where replacing or recharging batteries is difficult, such as implanted medical sensors or distributed IoT devices.

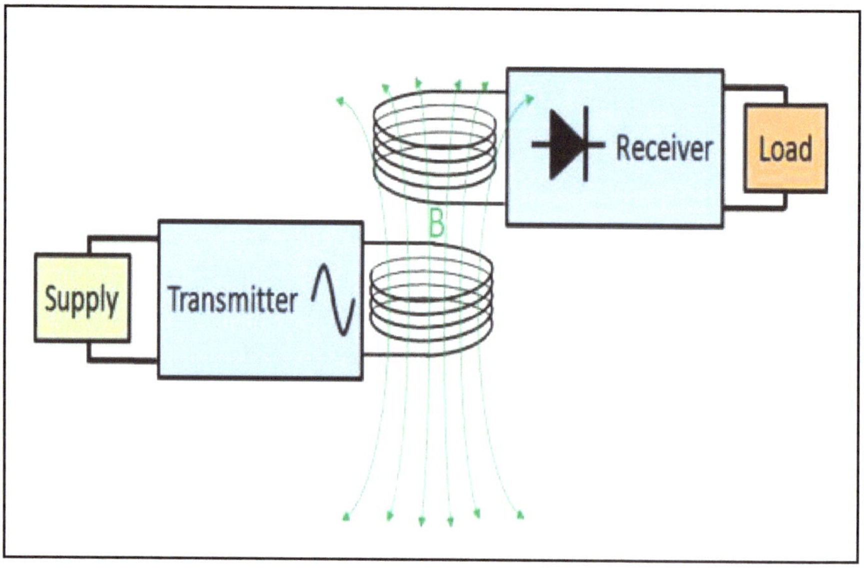

Wireless Power Transfer Circuit Working

9.4.2 Role of Metamaterials in Energy Transmission

The role of metamaterials in wireless power transfer is based on their ability to manipulate the propagation of electromagnetic waves. Conventional materials are bound by natural limits of wave refraction and focusing, but metamaterials can achieve negative permittivity or permeability, enabling phenomena such as negative refraction and field concentration. In WPT systems, metamaterials act as lenses or resonators that capture and direct energy toward a receiver with minimal loss. For example, a metamaterial slab placed between a transmitter and receiver coil can enhance magnetic coupling, improving efficiency even when the coils are misaligned or separated by greater distances. This enhancement reduces energy leakage into the surrounding environment and increases system robustness.

9.4.3 Resonant Coupling and Metamaterials

Resonant coupling is central to wireless power transfer, as it allows energy to be transferred efficiently between two devices operating at the same resonant frequency. Metamaterials enhance this process by creating strong local resonances that amplify the electromagnetic field. Split-ring resonators and wire arrays, common metamaterial structures, can be designed to resonate with specific frequencies, thereby intensifying the transfer pathway. This capability is crucial in medical applications where implants require steady power supply without frequent surgical interventions. By tuning resonant metamaterial structures, energy can be transmitted deeper into the body with minimal attenuation, opening possibilities for wireless recharging of pacemakers, neural stimulators, or drug delivery systems.

9.4.4 Energy Harvesting from Ambient Sources

Energy harvesting refers to the process of capturing and storing energy from surrounding sources such as sunlight, radiofrequency waves, vibrations, or thermal gradients. Metamaterials significantly enhance the ability to harvest ambient electromagnetic energy by concentrating fields and optimizing absorption at specific frequencies. Conventional antennas often struggle to efficiently capture weak, diffuse signals. Metamaterial antennas and absorbers, however, can be designed to resonate precisely with ambient frequencies such as Wi-Fi, cellular, or television signals, converting them into usable electrical power. This advancement makes self-powered sensors and wearable devices feasible, reducing dependence on batteries and external recharging.

9.4.5 Plasmonic and Photonic Energy Harvesting

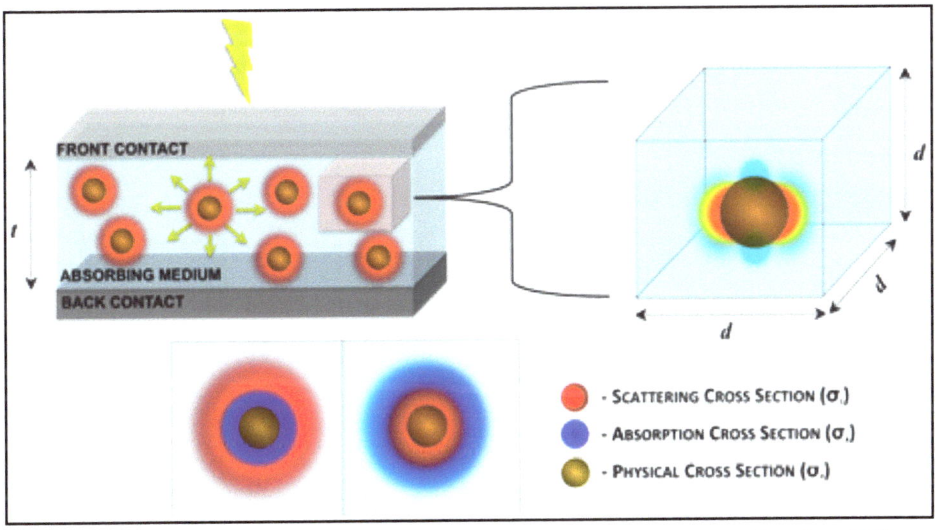

Plasmonic Nanoparticle Enhancement of Solution-Processed Solar Cells

In the optical domain, plasmonic metamaterials play a pivotal role in harvesting solar and ambient light energy. Plasmonic nanostructures can trap and guide light at subwavelength scales, concentrating energy into tiny regions for enhanced conversion. When integrated into photovoltaic systems, these metamaterials improve light absorption, thereby increasing the efficiency of solar cells. In addition, photonic metamaterials can create band gaps that filter and guide specific wavelengths of light, enabling highly selective energy harvesting. These designs are especially promising for wearable photovoltaics, where flexible and lightweight metamaterial coatings can be embedded into textiles or devices, creating energy-autonomous systems for healthcare monitoring or portable electronics.

9.4.6 Wireless Power in Medical Applications

One of the most impactful areas for metamaterial-enabled WPT is in medicine. Implanted medical devices such as pacemakers, cochlear implants, and biosensors traditionally rely on batteries that require replacement through invasive procedures. Metamaterials provide a safer and more efficient way to wirelessly deliver power through the skin and tissues.

By focusing electromagnetic energy directly onto the implant, metamaterial-assisted WPT reduces energy loss and heating of surrounding tissues, thereby improving patient safety. Additionally, metamaterial-based energy harvesters can power biosensors that continuously monitor glucose, oxygen levels, or neural activity without requiring external charging. This has the potential to transform long-term patient care by reducing surgical risks and improving device reliability.

9.4.7 Integration with the Internet of Things

The Internet of Things relies on billions of small, distributed devices that sense, communicate, and operate autonomously. Powering this vast network presents a significant challenge, as replacing or recharging batteries on such a scale is impractical.

Metamaterial-assisted energy harvesting offers a solution by capturing ambient electromagnetic signals, such as those from Wi-Fi or mobile communication systems, and converting them into usable power. Combined with wireless power transfer hubs, metamaterials create a sustainable power ecosystem for IoT devices. This integration not only enhances convenience but also reduces electronic waste, aligning with global goals for sustainability and efficiency.

9.4.8 Challenges in Wireless Power and Energy Harvesting

Despite their promise, metamaterial-based wireless power and energy harvesting systems face several challenges. Fabrication of metamaterials with nanoscale precision is complex and costly, particularly for large-scale applications. Efficiency losses can still occur due to environmental interference, frequency mismatch, or material imperfections. In biomedical applications, ensuring biocompatibility and safety is critical, as excessive energy absorption could damage tissues. Another challenge lies in regulatory approval, as transmitting power wirelessly at significant levels requires strict compliance with health and safety standards. Furthermore, scaling down metamaterial devices for flexible or wearable applications requires innovations in material science and integration with polymers or textiles.

9.5 QUANTUM METAMATERIALS AND QUANTUM LIGHT CONTROL

9.5.1 Concept of Quantum Metamaterials

Quantum metamaterials differ fundamentally from classical metamaterials in their operating principles. While conventional metamaterials rely on artificially structured elements to influence electromagnetic waves at classical scales, quantum metamaterials exploit quantum coherence, entanglement, and superposition to control wave–matter interactions.

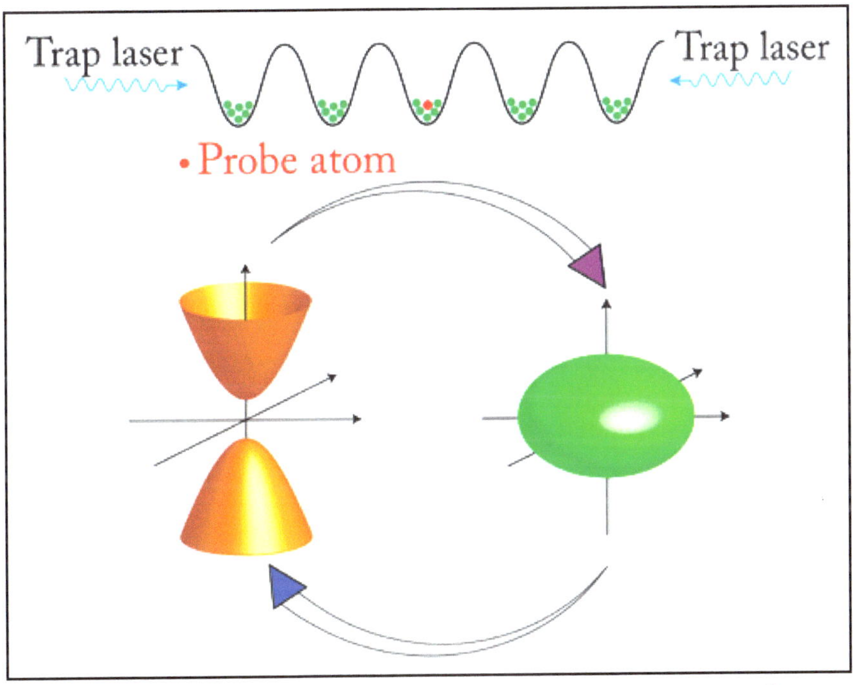

Fig: Major Leap towards a Perfect Quantum Metamaterial

This means that the properties of the material can be tuned not just by geometry and structure but also by quantum mechanical effects within its constituent elements. In many designs, quantum metamaterials are constructed from ensembles of quantum systems such as superconducting qubits, quantum dots, or ultracold atoms arranged in engineered arrays. These elements interact with light or electromagnetic fields in ways that classical oscillators cannot, enabling effects such as quantum superlensing, entanglement-assisted transmission, and single-photon control.

9.5.2 Quantum Light and its Importance

Quantum light refers to light that exhibits non-classical properties, including single-photon emission, squeezed states, or entangled photon pairs. Unlike classical light, which can be described by Maxwell's equations alone, quantum light requires the framework of quantum electrodynamics. Controlling quantum light is crucial for emerging quantum technologies.

In quantum computing, single photons act as qubits or carriers of information. In secure quantum communication, entangled photons enable protocols such as quantum key distribution that are resistant to eavesdropping. Metamaterials designed to manipulate quantum light can control its polarization, phase, or path at subwavelength scales, enabling compact and tunable quantum optical devices that surpass traditional optics.

9.5.3 Superconducting Qubits and Quantum Metamaterials

One of the most advanced platforms for realizing quantum metamaterials is superconducting qubits arranged in artificial lattices. These systems operate at cryogenic temperatures and exhibit coherence times sufficient to allow collective quantum behavior. When coupled to microwave photons in superconducting circuits, these arrays behave as engineered quantum media with tunable properties. For example, they can act as quantum mirrors, reflectors, or even artificial atoms that interact collectively with light. Such systems enable precise control of microwave quantum light, which is essential for building scalable quantum computers. By engineering the arrangement and coupling of qubits, researchers can design metamaterials that control photon transport in ways impossible with classical materials.

9.5.4 Quantum Dots and Nanophotonic Structures

Another promising approach involves integrating quantum dots into nanophotonic metamaterials. Quantum dots are semiconductor nanostructures that act as artificial atoms, emitting single photons or entangled photon pairs when excited. Embedding them in carefully designed photonic crystals or plasmonic metamaterials allows precise control over their emission properties. For instance, quantum dot–based metamaterials can enhance spontaneous emission through the Purcell effect or suppress it by creating band gaps. This ability to engineer the quantum emission of light is crucial for quantum communication systems and photonic quantum computing, where reliable single-photon sources are a fundamental requirement.

9.5.5 Plasmonic Quantum Metamaterials

Plasmonic metamaterials, which exploit the collective oscillation of electrons at metal surfaces, are also advancing quantum light control. At the nanoscale, plasmonic modes can couple strongly with quantum emitters such as molecules or quantum dots, leading to hybrid quantum states known as plasmon-exciton polaritons.

These hybrid states provide a platform for manipulating quantum light at room temperature, which is a significant advantage over superconducting or ultracold atom systems. Plasmonic quantum metamaterials can guide single photons along nanoscale pathways, confine light to volumes much smaller than the diffraction limit, and enable ultrafast quantum switches. Such systems are key for integrating quantum devices into compact, on-chip platforms for quantum information processing.

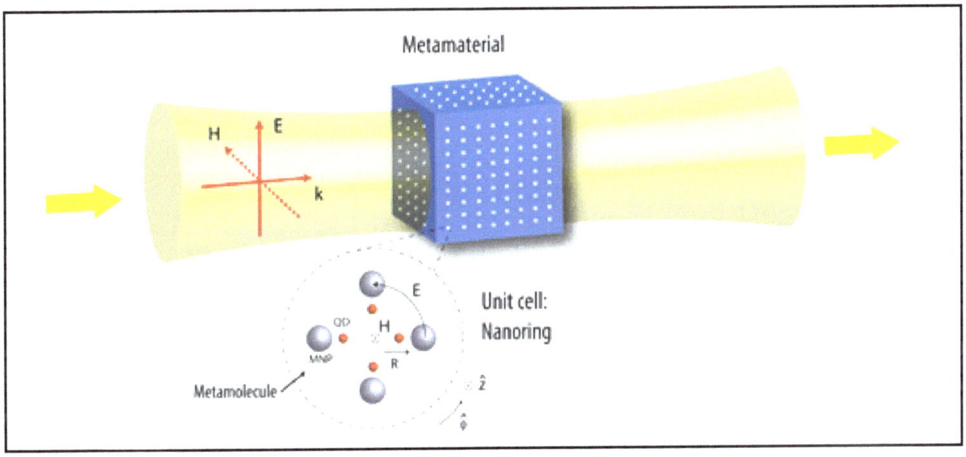

Fig: Schematic of the Quantum Plasmonic Metamaterial

9.5.6 Quantum Coherence and Collective Effects

A defining feature of quantum metamaterials is their reliance on collective quantum coherence. When many quantum emitters or qubits interact coherently, they can produce collective phenomena such as superradiance, subradiance, or entangled emission. These effects amplify or suppress light emission depending on the arrangement and phase of the emitters.

By designing the spatial and spectral properties of metamaterials, engineers can harness these collective effects to build devices with unprecedented performance. For example, superradiant quantum metamaterials can generate intense, coherent bursts of light, while subradiant systems can store quantum information in long-lived collective states. This interplay between structure and quantum coherence defines the unique power of quantum metamaterials.

9.5.7 Quantum Light Control for Communication

Quantum communication relies on transmitting information using the quantum states of photons. Quantum metamaterials provide a platform for encoding, transmitting, and decoding quantum information more efficiently. By manipulating phase, polarization, and entanglement at the nanoscale, metamaterials can function as compact quantum routers, filters, and modulators. For instance, metasurfaces designed at the quantum level can redirect entangled photon pairs along specific pathways without destroying their correlations. This capability is essential for building secure quantum communication networks, where photons must maintain their quantum states over long distances despite environmental noise. Quantum metamaterials thus form a backbone for future quantum internet infrastructure.

9.5.8 Applications in Quantum Sensing

Quantum metamaterials also hold promise for next-generation sensing technologies. Quantum sensors exploit phenomena such as superposition and entanglement to achieve sensitivities far beyond classical limits. Metamaterials can enhance these effects by concentrating fields or isolating quantum states from environmental noise. For example, metamaterial-enhanced quantum magnetometers can detect minute changes in magnetic fields with applications ranging from brain imaging to navigation. Similarly, quantum optical metamaterials can detect extremely low concentrations of biomolecules by monitoring shifts in quantum resonance. These advancements make quantum metamaterials valuable not only for fundamental science but also for applied medical and industrial sensing.

CHAPTER 10

Challenges and Future Perspectives

10.1 SCALABILITY AND MASS PRODUCTION

10.1.1 Challenges in Scaling Metamaterial Fabrication

The fundamental challenge in scaling metamaterials lies in the precision required for their subwavelength structures. The performance of a metamaterial depends critically on the geometry, periodicity, and alignment of its unit cells. At optical, terahertz, or microwave frequencies, these structures can range from tens of nanometers to millimeters. Traditional fabrication techniques, such as electron-beam lithography or focused ion beam milling, offer the required resolution but are slow and expensive, making them impractical for large-scale production. In addition, many of these processes are limited to rigid substrates, which restricts their integration into flexible or wearable applications. Achieving uniformity across large areas is another hurdle, as even small deviations in structure can drastically affect electromagnetic, acoustic, or thermal properties.

10.1.2 Advances in Lithography for Mass Production

Lithographic techniques have been adapted to address scalability challenges. Photolithography, commonly used in semiconductor manufacturing, allows for the replication of patterns over wafer-scale areas. Innovations such as deep ultraviolet (DUV) and extreme ultraviolet (EUV) lithography have pushed the resolution limits, enabling nanoscale patterning necessary for optical and plasmonic metamaterials. Nanoimprint lithography (NIL) offers a cost-effective alternative by using a master mold to stamp patterns onto substrates repeatedly. These methods enable large-area fabrication with high reproducibility, which is critical for commercial applications such as sensors, antennas, or optical devices. However, the cost of master molds, equipment, and alignment precision still poses challenges for ultra-large-scale deployment.

10.1.3 3D Printing and Additive Manufacturing

Additive manufacturing, commonly known as 3D printing, has emerged as a versatile tool for scalable metamaterial production. Unlike lithography, which is largely planar, 3D printing allows the fabrication of complex three-dimensional unit cells required for acoustic, mechanical, or thermal metamaterials. Techniques such as two-photon polymerization, direct laser writing, and micro-stereolithography provide submicron resolution, making it possible to realize intricate geometries with high fidelity. Moreover, additive manufacturing enables rapid prototyping and iterative design, accelerating research and development cycles. For larger-scale industrial applications, continuous and high-speed 3D printing processes are being developed, potentially allowing the production of meter-scale metamaterial panels for aerospace, architecture, or energy applications.

10.1.4 Roll-to-Roll and Flexible Manufacturing

For applications requiring large-area and flexible metamaterials, roll-to-roll manufacturing techniques offer a promising solution. In these processes, metamaterial patterns are continuously printed or embossed onto flexible substrates such as polymers or thin metal films. This approach is especially relevant for wearable electronics, flexible antennas, solar energy harvesting devices, and thin-film optics. Roll-to-roll processes are compatible with nanoimprint lithography, laser patterning, or even inkjet printing of functional materials, enabling high-throughput production. Challenges remain in achieving nanoscale precision and consistent layer thickness over long rolls, but ongoing research in materials and process control is gradually overcoming these limitations.

10.1.5 Material Considerations for Scalable Production

The choice of materials plays a crucial role in the scalability of metamaterials. Polymers, metals, ceramics, and composites each present unique advantages and limitations. Polymers are lightweight, flexible, and easy to process, making them ideal for large-area and roll-to-roll fabrication, but they may exhibit higher losses at high frequencies. Metals, such as copper, silver, or gold, provide excellent electromagnetic properties but can be expensive and challenging to pattern at nanoscale. Ceramics offer high thermal and mechanical stability but often require high-temperature processing that limits substrate choice. Composites, which combine different materials, allow tailored properties but introduce complexity in fabrication. Selecting the appropriate material depends on balancing performance, manufacturability, and cost for the intended application.

10.1.6 Standardization and Design for Manufacturability

Another critical aspect of scalability is standardization. Metamaterial unit cells and device architectures must be designed for manufacturability to reduce costs and improve yield. Simplifying unit cell geometries, using modular designs, and employing self-assembly techniques can enhance production efficiency. Standardization also facilitates integration into existing industrial processes and promotes interoperability in commercial systems. Furthermore, simulation tools that predict fabrication tolerances and performance variations can guide the design of robust metamaterials that retain their functionality despite minor imperfections.

10.1.7 Automation and Quality Control

Automation and quality control are essential for mass production of metamaterials. Advanced imaging and metrology systems can monitor nanoscale patterns in real time, ensuring consistent performance across large batches. Machine learning algorithms are increasingly being applied to detect defects, optimize fabrication parameters, and predict performance outcomes. Automation not only reduces production costs but also improves reproducibility, which is critical for applications in aerospace, defense, and medical devices, where failure can have serious consequences.

10.1.8 Economic Considerations and Industrial Adoption

While technological advancements are critical, economic feasibility ultimately determines the commercial success of metamaterials. Large-scale production must balance performance with cost-efficiency. Industries such as telecommunications, aerospace, and consumer electronics are particularly sensitive to cost per unit area or volume. Scaling production through methods like roll-to-roll printing or 3D additive manufacturing can reduce costs, but initial capital investment and process development remain significant barriers. Strategic partnerships between research institutions and industry are essential to translate laboratory-scale innovations into commercially viable metamaterials.

10.1.9 Future Directions

The future of scalable metamaterial production is likely to be shaped by advances in hybrid fabrication techniques, new material systems, and automated design-to-manufacture pipelines. Emerging approaches such as self-assembled nanostructures, tunable metamaterials, and multifunctional composites will enable large-area, high-performance devices with reduced fabrication complexity. Integration of machine learning and AI-driven process optimization will further enhance scalability and reproducibility. In the long term, scalable metamaterials could enable widespread deployment in consumer electronics, smart textiles, energy systems, medical devices, and defense technologies, transforming both industry and daily life.

10.2 LOSSES, BANDWIDTH, AND MATERIAL LIMITATIONS

Losses, bandwidth, and material limitations define the practical boundaries of metamaterial performance. Material absorption, geometric imperfections, and resonant constraints impose significant challenges that must be addressed for real-world applications. While these limitations are substantial, ongoing research in low-loss materials, hybrid structures, active compensation, and tunable designs is gradually overcoming them. By carefully balancing loss reduction, bandwidth optimization, and material selection, engineers can create metamaterials that realize their full potential across diverse technological domains. The continued evolution of materials science, fabrication techniques, and design strategies promises a future where metamaterials are both high-performance and scalable, enabling transformative applications in optics, acoustics, thermal management, and beyond.

10.2.1 Understanding Material Losses

Losses in metamaterials refer to the dissipation of energy that occurs when electromagnetic or acoustic waves interact with the metamaterial structure. In electromagnetic metamaterials, these losses often arise from the inherent resistive properties of metals, dielectric absorption, or scattering at interfaces. In optical and terahertz metamaterials, metallic components such as gold or silver introduce Ohmic losses that reduce efficiency. Similarly, in acoustic or mechanical metamaterials, internal friction and viscoelastic damping contribute to energy dissipation. Material losses directly impact device performance by attenuating signals, reducing field enhancement, and limiting the effectiveness of phenomena such as superlensing or cloaking. In high-precision applications like sensing or quantum light control, even minor losses can degrade signal-to-noise ratios and diminish functionality.

10.2.2 Ohmic and Dielectric Losses in Electromagnetic Metamaterials

Electromagnetic metamaterials often rely on metallic inclusions, such as split-ring resonators or wire arrays, embedded in dielectric substrates. Ohmic losses in metals arise from the finite conductivity of electrons, causing heat generation and attenuation of the electromagnetic field. Dielectric losses occur when the substrate material absorbs part of the energy from the oscillating electric field, converting it into heat. These combined losses not only limit the efficiency of energy transfer and field enhancement but also constrain the scaling of metamaterial devices for high-power applications. Engineers must carefully balance geometric design, choice of metals, and dielectric properties to minimize losses while maintaining the desired metamaterial response.

10.2.3 Bandwidth Limitations

Bandwidth refers to the frequency range over which a metamaterial exhibits its engineered properties. Many metamaterials, particularly those relying on resonant structures, exhibit inherently narrow bandwidths. For instance, split-ring resonators have sharp resonant frequencies that produce strong negative permeability or permittivity only over a limited spectral range. While resonance enhances certain effects, it restricts applications requiring broadband performance, such as wide-spectrum imaging, high-speed communication, or energy harvesting from multiple frequency sources. Bandwidth limitations are further exacerbated by material dispersion, where the refractive index varies with frequency, leading to deviations from the desired response outside the resonant range. Overcoming these limitations requires innovative designs, including multi-resonant structures, gradient-index metamaterials, or tunable systems that dynamically adjust their response.

10.2.4 Trade-Offs between Loss and Bandwidth

Losses and bandwidth are often interrelated in metamaterial design. Structures that achieve strong resonant responses typically concentrate electromagnetic or acoustic energy in small regions, amplifying material losses. Conversely, attempts to broaden bandwidth by reducing resonant quality factors can weaken the metamaterial effect, diminishing phenomena such as negative refraction or superlensing. This trade-off poses a fundamental engineering challenge: achieving high-performance metamaterials that maintain both low losses and sufficient bandwidth. Researchers are exploring solutions such as hybrid designs, combining low-loss dielectrics with metallic resonators, or employing active and tunable elements to compensate for losses dynamically.

10.2.5 Limitations Imposed by Material Properties

The choice of constituent materials imposes intrinsic limitations on metamaterial performance. Metals used in electromagnetic metamaterials suffer from finite conductivity and skin depth effects, particularly at optical frequencies, leading to significant absorption losses. Dielectric materials, while low-loss in some regimes, may not sustain high field intensities without breakdown or heating. In acoustic metamaterials, the mechanical properties of the base material, such as elasticity and density, limit the achievable control over sound propagation and impedance matching. Thermal metamaterials are constrained by the heat capacity and thermal conductivity of the materials used, which restrict the ability to create strong gradients or cloaking effects. Composite or hybrid materials can mitigate some limitations, but introducing multiple materials often complicates fabrication and may introduce interfacial losses or instability.

10.2.6 Challenges in Scaling to High Frequencies

As metamaterials are scaled to operate at higher frequencies, such as terahertz or visible light, material limitations become more pronounced. Metals that function efficiently at microwave frequencies exhibit increased losses in the optical range due to electron scattering and interband transitions. Similarly, dielectric losses often rise at higher frequencies due to absorption resonances. Nanoscale fabrication introduces additional challenges, as surface roughness, grain boundaries, and defects can exacerbate losses and limit device performance. Acoustic and mechanical metamaterials face similar scaling issues, where miniaturization can lead to unwanted resonances, reduced structural stability, and increased dissipation. Overcoming these high-frequency limitations requires advanced materials with low-loss properties, precision nanofabrication techniques, and design strategies that distribute energy more efficiently.

10.2.7 Active and Tunable Metamaterials as a Solution

One promising approach to mitigate losses and bandwidth limitations is the development of active or tunable metamaterials. By incorporating gain media, such as semiconductor amplifiers or optical pumps, metamaterials can compensate for inherent losses, effectively restoring signal strength and enhancing performance. Tunable metamaterials, employing MEMS, varactors, or phase-change materials, allow dynamic adjustment of resonance frequencies and material response, enabling broader bandwidth operation and adaptive functionality. These strategies not only address energy dissipation and narrow-band limitations but also expand the versatility of metamaterials in real-world applications, such as adaptive lenses, reconfigurable antennas, and smart acoustic devices.

10.2.8 Future Perspectives

Addressing losses, bandwidth, and material limitations remains central to the continued advancement of metamaterials. Emerging materials, such as two-dimensional graphene, transition metal dichalcogenides, and low-loss ceramics, offer new possibilities for high-frequency, low-dissipation designs. Innovations in fabrication, including nanoimprinting, self-assembly, and hybrid additive manufacturing, are reducing defects and enabling more precise structures. The integration of active components and tunable systems promises adaptive metamaterials capable of maintaining optimal performance across broader spectral and operational ranges. As these developments mature, metamaterials will transition from experimental demonstrations to practical, scalable devices in telecommunications, energy harvesting, biomedical imaging, and defense systems.

10.3 INTERDISCIPLINARY RESEARCH NEEDS

Interdisciplinary research is not merely beneficial but essential for the continued advancement of metamaterials. The complexity of design, fabrication, characterization, and integration demands expertise spanning physics, materials science, engineering, nanotechnology, computation, and biomedicine. Collaborative approaches allow researchers to overcome challenges related to losses, bandwidth, scalability, and functionality, enabling the translation of laboratory innovations into real-world applications. Education, shared research platforms, and integrated funding models are key to nurturing a workforce capable of driving these interdisciplinary efforts.

10.3.1 Physics and Electromagnetic Theory

Fundamental physics underpins the behavior of metamaterials, particularly in controlling waves and fields. Understanding electromagnetic interactions, acoustic vibrations, and thermal conduction at subwavelength scales is essential to designing functional metamaterials. Classical Maxwell's equations are extended with effective medium theories to predict the collective behavior of engineered unit cells. Quantum mechanics becomes critical when dealing with nanoscale or quantum metamaterials, where effects such as superposition, entanglement, and coherence influence material response. Insights from solid-state physics inform the understanding of plasmonic resonances, phonon interactions, and energy dissipation mechanisms. Collaboration between theorists and experimentalists ensures that the predicted phenomena are accurately translated into tangible structures, bridging abstract physics with practical device design.

10.3.2 Material Science and Chemistry

The realization of metamaterials relies on advances in material science and chemistry. Choice of materials—metals, polymers, ceramics, composites, or emerging 2D materials—affects electromagnetic, acoustic, thermal, and mechanical responses. Material properties such as conductivity, permittivity, permeability, elasticity, and thermal stability directly influence the performance and feasibility of metamaterials. Chemistry provides methods for synthesizing nanostructures, functionalizing surfaces, and creating hybrid composites with tailored properties. Material scientists must work closely with engineers to ensure that the chosen substances are compatible with fabrication processes, scalable production, and intended operating environments. Developing low-loss, high-performance, and biocompatible materials often requires novel chemical formulations, molecular engineering, and surface treatments.

10.3.3 Nanotechnology and Fabrication Techniques

Fabrication is a major bottleneck in metamaterial research, particularly at optical and terahertz frequencies where nanoscale precision is required. Nanotechnology provides the tools for lithography, electron-beam patterning, nanoimprinting, self-assembly, and 3D printing, enabling the creation of intricate subwavelength structures. Innovations in additive manufacturing and roll-to-roll processing extend these techniques to larger areas and flexible substrates. Collaboration between nanotechnologists and designers ensures that the fabricated structures faithfully reproduce the intended geometry and performance. Additionally, interdisciplinary work with materials scientists addresses challenges such as adhesion, surface roughness, and defect management, which can significantly influence metamaterial functionality.

10.3.4 Computational Modeling and Simulation

Metamaterials often involve complex geometries and multi-physical interactions that are difficult to analyze analytically. Computational modeling and numerical simulation are essential to predict behavior, optimize design, and reduce costly experimental iterations. Finite element methods, finite-difference time-domain simulations, and multiphysics modeling allow researchers to explore electromagnetic, acoustic, and thermal responses in silico.

Interdisciplinary collaboration with computer scientists and applied mathematicians is crucial to develop efficient algorithms, handle large-scale simulations, and implement optimization techniques such as machine learning. These computational tools accelerate the translation of theoretical designs into practical devices by predicting losses, resonance frequencies, field distributions, and performance limits.

10.3.5 Electrical, Mechanical, and Optical Engineering

Engineering disciplines contribute to the practical integration of metamaterials into functional devices. Electrical engineers focus on RF, microwave, terahertz, and optical applications, designing antennas, filters, and waveguides with metamaterial inclusions. Mechanical engineers study acoustic, vibration, and structural metamaterials, optimizing energy absorption, damping, and mechanical stability. Optical engineers work on lenses, imaging systems, and photonic devices, ensuring precise control of light propagation. Effective collaboration among these fields ensures that metamaterials not only demonstrate exceptional properties in isolation but also integrate seamlessly into real-world systems, meeting requirements such as durability, reliability, and environmental tolerance.

10.3.6 Biomedical and Life Sciences

Interdisciplinary research extends into biomedical and life sciences, where metamaterials are applied in imaging, sensing, and therapeutic devices. Biomedical engineers and biophysicists work with materials scientists to develop biocompatible metamaterials for implantable sensors, MRI enhancement, and biosensing platforms. Understanding tissue interactions, cellular response, and biofluids is essential for designing safe and effective devices. Collaboration with chemists enables functionalization for selective biomolecule detection, while electrical and optical engineers ensure that signals can be accurately transmitted and received. This integration of life sciences and engineering accelerates translational research, bringing metamaterials from laboratory prototypes to clinical applications.

10.3.7 Interdisciplinary Challenges and Collaboration

Despite clear benefits, interdisciplinary research in metamaterials faces several challenges. Differences in terminology, methodology, and research culture across fields can create communication barriers. Integrating theoretical modeling with experimental fabrication requires constant feedback loops and mutual understanding. Scaling metamaterials for industrial use necessitates collaboration between academia and industry, requiring coordination across research, manufacturing, and regulatory teams. Funding structures and publication systems may also favor single-discipline work, making interdisciplinary collaboration more complex. Overcoming these barriers requires intentional collaboration frameworks, shared research goals, and platforms for cross-disciplinary knowledge exchange.

10.3.8 Education and Workforce Development

The successful development of metamaterials depends not only on research but also on cultivating an interdisciplinary workforce. Educational programs that integrate physics, materials science, engineering, nanotechnology, and computational modeling are essential to train researchers capable of bridging these domains. Workshops, joint research centers, and collaborative laboratories foster hands-on experience with both theoretical and experimental tools. Mentorship and cross-disciplinary coursework prepare scientists and engineers to address the multifaceted challenges of metamaterials, from design and fabrication to testing and commercialization. A workforce with broad expertise accelerates innovation and ensures that metamaterials can be translated into practical technologies efficiently.

10.3.9 Future Directions

The future of metamaterials is inherently interdisciplinary. Advancements in tunable, active, and quantum metamaterials will require collaboration across multiple domains, including photonics, nanotechnology, quantum mechanics, and biomedicine. Emerging areas such as soft metamaterials, smart textiles, and energy-harvesting systems demand integrated expertise in flexible materials, electronics, and system-level design. Artificial intelligence and machine learning will further enhance design optimization, necessitating collaboration with computer scientists and data analysts. By fostering strong interdisciplinary research networks, the metamaterials community can accelerate the pace of innovation, address existing limitations, and expand applications across multiple sectors.

10.4 SUSTAINABILITY AND COST CONSIDERATIONS

Sustainability and cost considerations are central to the successful deployment of metamaterials across industries. Material selection, fabrication processes, energy consumption, waste management, and economic feasibility all influence whether metamaterials can move beyond laboratory prototypes to practical applications. By focusing on low-impact materials, scalable and efficient production techniques, recycling-friendly designs, and life-cycle assessment, researchers and manufacturers can reduce environmental and economic barriers. Balancing performance, cost, and sustainability requires interdisciplinary collaboration, innovation, and careful planning. As advances continue in materials science, nanotechnology, and fabrication methods, metamaterials are poised to become not only technologically transformative but also environmentally responsible and economically viable, supporting a sustainable and high-performance future.

10.4.1 Material Selection and Environmental Impact

The sustainability of metamaterials begins with material selection. Traditional metamaterials often rely on metals such as gold, silver, or copper for electromagnetic functionality, or high-performance polymers, ceramics, and composites for structural and acoustic applications. Metals like gold and silver offer excellent electrical and plasmonic properties but are expensive and have significant environmental footprints associated with mining, refining, and disposal.

Polymers and composites, while cheaper and lightweight, may involve petroleum-based precursors and produce microplastic waste, raising environmental concerns. Ceramics, although chemically stable, require energy-intensive high-temperature processing. To improve sustainability, researchers are exploring alternatives such as biodegradable polymers, low-loss aluminum or copper alloys, recycled materials, and 2D materials like graphene that offer high performance with lower environmental impact. Material choice must balance functionality with lifecycle sustainability, ensuring that metamaterial production does not create disproportionate environmental burdens.

10.4.2 Fabrication Techniques and Energy Consumption

The production of metamaterials often requires precision fabrication techniques, including electron-beam lithography, nanoimprinting, 3D printing, and additive manufacturing. While these methods provide the resolution needed to achieve subwavelength structures, they can be energy-intensive, time-consuming, and costly.

Electron-beam lithography, for instance, offers nanoscale precision but consumes large amounts of energy and operates slowly, limiting throughput. Similarly, some 3D printing methods require high-intensity lasers or long curing cycles, which contribute to high energy usage. Roll-to-roll manufacturing and other scalable techniques reduce energy and material waste by enabling continuous production over large areas, but achieving nanoscale fidelity remains challenging. Optimizing fabrication processes to reduce energy consumption, minimize waste, and improve throughput is therefore a central consideration in sustainable metamaterial design.

10.4.3 Cost of Raw Materials and Processing

Economic feasibility is a major barrier to the commercialization of metamaterials. High-quality metals, specialized polymers, and composite materials often account for a substantial portion of production costs. Additionally, the fabrication infrastructure, including clean rooms, lithography equipment, and precision 3D printers, requires significant capital investment. The combination of expensive materials and complex processing leads to high unit costs, limiting the adoption of metamaterials in cost-sensitive applications such as consumer electronics or large-area energy devices. Researchers are investigating cost-reduction strategies, including the use of inexpensive metals, polymer composites, and hybrid materials, as well as simplified geometries that retain functional performance while reducing fabrication complexity. Standardization and modular design approaches also help lower costs by enabling batch fabrication and easier integration into existing industrial processes.

10.4.4 Waste Management and Recycling

Sustainability considerations extend beyond production to include end-of-life management and recycling. Many metamaterials are designed for specialized high-performance applications, making recycling or reprocessing challenging. Metallic metamaterials, for instance, may contain multiple layers or coatings that complicate separation and recovery. Polymers and composites can produce hazardous byproducts if incinerated or improperly disposed of. Implementing recycling-friendly designs, such as single-material layers or separable structures, can reduce environmental impact and support circular material lifecycles. Researchers are also exploring biodegradable or recyclable materials, particularly for disposable or wearable metamaterials, to reduce long-term environmental burdens. Waste management strategies must be integrated into both the design and manufacturing stages to ensure that metamaterials are truly sustainable throughout their lifecycle.

10.4.5 Life-Cycle Assessment and Sustainability Metrics

To evaluate the sustainability of metamaterials, life-cycle assessment (LCA) methods are increasingly applied. LCA considers the environmental impact of materials and processes from raw material extraction through production, usage, and disposal. Metrics such as energy consumption, carbon footprint, water usage, and toxic emissions help quantify the sustainability of different metamaterial designs. By comparing alternative materials, fabrication techniques, and operational scenarios, designers can make informed choices that optimize performance while minimizing environmental costs. This approach encourages the adoption of eco-friendly practices, guides material selection, and informs regulatory compliance for industrial applications.

10.4.6 Balancing Performance and Cost Efficiency

Sustainability and cost considerations often intersect with performance requirements. High-performance metamaterials may demand expensive or rare materials and precision fabrication, while more sustainable or low-cost options may compromise functionality. Achieving an optimal balance requires interdisciplinary collaboration between materials scientists, engineers, and economists.

Strategies such as hybrid materials, tunable designs, and additive manufacturing can enhance performance without significantly increasing costs or environmental impact. Similarly, computational modeling and machine learning can predict the trade-offs between material choice, geometry, and functionality, allowing designers to optimize both performance and sustainability simultaneously.

10.4.7 Industrial and Commercial Implications

Sustainable and cost-effective metamaterials are critical for industrial and commercial adoption. In sectors such as telecommunications, aerospace, energy harvesting, and wearable electronics, large-scale production is necessary to achieve economic viability. Companies are increasingly seeking materials that can be produced reliably at scale while minimizing environmental and energy costs.

Regulatory requirements for environmental compliance, energy efficiency, and waste management further emphasize the importance of sustainable metamaterial design. Early integration of sustainability considerations in research and development accelerates commercialization and aligns with global trends toward green technologies and circular economy practices.

10.4.8 Future Directions

The future of sustainable and cost-effective metamaterials depends on innovations in materials, fabrication, and system design. Emerging approaches include the use of low-cost nanocomposites, 2D materials such as graphene and transition metal dichalcogenides, and self-assembled nanostructures that reduce fabrication complexity. Advances in roll-to-roll nanoimprinting, additive manufacturing, and scalable lithography will lower energy consumption and waste, enabling larger-area production. Integration of machine learning for design optimization and predictive modeling will further reduce material usage while improving performance. Additionally, interdisciplinary collaborations between scientists, engineers, and economists will support the development of metamaterials that meet both functional and sustainability criteria, facilitating widespread adoption.

10.5 FUTURE DIRECTIONS: 4D AND BIO-INSPIRED METAMATERIALS

4D and bio-inspired metamaterials represent the next frontier in engineered materials, combining dynamic adaptability with nature-inspired efficiency. By incorporating time-dependent responsiveness and hierarchical, multifunctional design principles, these materials offer unprecedented control over waves, energy, and mechanical forces. While challenges in fabrication, material selection, and scalability remain, continued interdisciplinary research promises to overcome these barriers. The integration of 4D and bio-inspired principles will enable metamaterials that are not only high-performing but also intelligent, adaptive, and environmentally compatible, opening transformative possibilities in technology and society.

10.5.1 4D Metamaterials

4D metamaterials represent an evolution of traditional three-dimensional structures by incorporating the fourth dimension: time. These materials are designed to change their properties or structure dynamically in response to external stimuli such as temperature, light, magnetic or electric fields, or mechanical stress.

By integrating responsive materials like shape-memory polymers, hydrogels, or phase-change materials into the metamaterial lattice, engineers can create structures that morph, self-assemble, or adapt their functionality on demand. This dynamic behavior allows metamaterials to operate across multiple frequencies, alter their wave propagation characteristics, or even reconfigure their mechanical properties in real time. For instance, a 4D acoustic metamaterial could shift its band gaps to selectively block or transmit different sound frequencies, while a thermal 4D metamaterial could redirect heat flow in response to environmental changes.

Mechanisms Enabling 4D Metamaterials

The realization of 4D metamaterials depends on embedding stimuli-responsive elements within the metamaterial architecture. Shape-memory polymers, which undergo reversible deformation when heated or cooled, can act as tunable structural components. Hydrogels expand or contract in response to humidity or pH changes, allowing programmable mechanical or acoustic responses.

Phase-change materials such as vanadium dioxide can alter their electromagnetic properties when transitioning between insulating and conductive states. Integration of microelectromechanical systems (MEMS) or nanoscale actuators further enables precise control over the metamaterial response. The combination of these materials and mechanisms allows for temporal modulation of metamaterial properties, effectively adding a "time dimension" to their functionality and significantly expanding their potential applications.

Applications of 4D Metamaterials

The adaptive nature of 4D metamaterials has wide-ranging implications. In aerospace and automotive engineering, 4D mechanical metamaterials can enable morphing surfaces or adaptive vibration control, improving aerodynamic efficiency and structural resilience. In optics and telecommunications, dynamically tunable metamaterials allow adaptive lenses, reconfigurable antennas, and modulators that optimize signal transmission across varying conditions.

In energy systems, 4D thermal metamaterials can redirect heat flow to enhance cooling or energy harvesting efficiency. The biomedical field can also benefit, where 4D metamaterials integrated into implants or prosthetics can adapt to physiological conditions, providing responsive mechanical support or controlled drug delivery. The dynamic adaptability of 4D metamaterials positions them as a key enabling technology for smart systems in multiple domains.

10.5.2 Bio-Inspired Metamaterials

Bio-inspired metamaterials derive their design principles from the efficiency, adaptability, and multifunctionality observed in natural systems. Nature provides abundant examples of hierarchical structures, anisotropic properties, and multifunctional surfaces that achieve remarkable performance with minimal material usage.

For example, the micro- and nanoscale structures of butterfly wings produce iridescence and selective light filtering, while the hierarchical lattice of bone provides lightweight mechanical strength. Bio-inspired metamaterials seek to replicate these strategies to create materials that are lightweight, strong, multifunctional, and environmentally adaptive. By mimicking biological forms and mechanisms, these metamaterials can achieve properties that are difficult or impossible with purely synthetic designs.

Mechanisms in Bio-Inspired Metamaterials

Bio-inspired designs often incorporate hierarchical structuring, gradient properties, and multifunctionality. Hierarchical structures, with features ranging from nanoscale to macroscale, enhance mechanical, thermal, and optical properties simultaneously. Gradient materials, inspired by structures such as bamboo or seashells, allow gradual transitions in stiffness, density, or refractive index, reducing stress concentrations and improving wave control. Surface functionalization mimics natural features like lotus leaves or shark skin, creating superhydrophobicity, self-cleaning surfaces, or drag reduction. Incorporating responsive or self-healing components further enhances adaptability, allowing bio-inspired metamaterials to react to environmental changes in ways similar to living organisms.

Applications of Bio-Inspired Metamaterials

Bio-inspired metamaterials are applicable across a wide range of fields. In structural engineering, lightweight yet strong hierarchical materials reduce weight while maintaining strength and resilience, critical for aerospace, automotive, and civil infrastructure. In optics and photonics, biomimetic surfaces can control light propagation, color generation, or polarization for advanced sensors and imaging devices. Acoustic metamaterials inspired by natural sound management, such as owl wings or termite mounds, offer enhanced noise reduction and vibration control. In biomedical applications, bio-inspired metamaterials can mimic tissue mechanics, creating implants or scaffolds with gradient elasticity that promotes cell growth and tissue integration. Energy applications benefit as well, where bio-inspired thermal and photonic metamaterials improve energy efficiency and light management.

10.5.3 Integration of 4D and Bio-Inspired Principles

The convergence of 4D and bio-inspired concepts represents a particularly promising frontier. Adaptive metamaterials inspired by natural structures can respond dynamically to environmental stimuli, combining the efficiency and multifunctionality of biological systems with the temporal tunability of 4D designs. For example, a bio-inspired 4D thermal metamaterial could dynamically adjust heat flow across a hierarchical lattice in response to temperature changes, mimicking thermoregulation mechanisms in animals. Similarly, adaptive optical metamaterials could reproduce the color-changing skin of cephalopods for camouflage or display technologies. This integration expands the functional possibilities, creating materials that are not only high-performing but also intelligent, adaptive, and environmentally compatible.

10.5.4 Challenges and Research Needs

Despite their potential, the development of 4D and bio-inspired metamaterials faces significant challenges. Fabrication at multiple scales, from nanoscale features to macroscale architectures, remains technically complex and resource-intensive. Material compatibility, especially for responsive or multifunctional systems, is difficult to achieve without sacrificing durability or efficiency. Computational design of adaptive and hierarchical structures requires advanced simulation techniques and interdisciplinary expertise in physics, materials science, biology, and engineering. Furthermore, translating laboratory-scale prototypes into scalable, cost-effective, and environmentally sustainable devices requires continued innovation in materials, manufacturing, and testing methodologies.

10.5.5 Future Outlook

The future of 4D and bio-inspired metamaterials is promising, driven by advances in responsive materials, additive manufacturing, nanofabrication, and computational modeling. Integration with artificial intelligence and machine learning will enable predictive design, optimization, and adaptive control, further enhancing functionality. Interdisciplinary collaboration across materials science, engineering, biology, and physics will accelerate innovation, enabling multifunctional metamaterials that are lightweight, energy-efficient, and environmentally sustainable. As fabrication techniques mature, these metamaterials are likely to transform fields such as aerospace, defense, healthcare, energy systems, and wearable technology, ushering in a new era of smart, adaptive materials.

CHAPTER 11

Case Studies and Real-World Implementations

11.1 CLOAKING DEVICES AND EXPERIMENTAL DEMONSTRATIONS

Cloaking devices represent one of the most captivating applications of metamaterials, translating theoretical principles into experimentally realizable systems. Early microwave demonstrations validated the concept, while optical, acoustic, thermal, and mechanical cloaks have expanded the scope and versatility of cloaking technologies.

Experimental achievements highlight both the promise and the challenges, including material losses, bandwidth limitations, fabrication precision, and scaling constraints. Ongoing research in hybrid designs, active cloaks, and multifunctional metamaterials is addressing these challenges, moving cloaking from laboratory experiments toward practical applications.

As fabrication techniques advance and interdisciplinary collaboration continues, metamaterial-based cloaking devices are poised to become an integral part of next-generation technologies in defense, imaging, energy, and acoustic management.

11.1.1 Principles of Cloaking

Cloaking relies on the principle of controlling wave propagation so that waves travel around an object as if it were not present. In electromagnetic cloaking, metamaterials can be engineered to produce a spatially varying permittivity and permeability, effectively bending light or other electromagnetic waves around a target. Acoustic cloaking uses analogous techniques, designing materials with tailored mass density and bulk modulus to redirect sound waves.

Thermal cloaking manipulates the effective thermal conductivity of materials to divert heat flow around an object. The common thread across these approaches is the use of engineered unit cells or structures to create a graded, anisotropic response that guides waves smoothly without scattering. The challenge lies in implementing these graded properties at the required scales while minimizing losses and maintaining broadband performance.

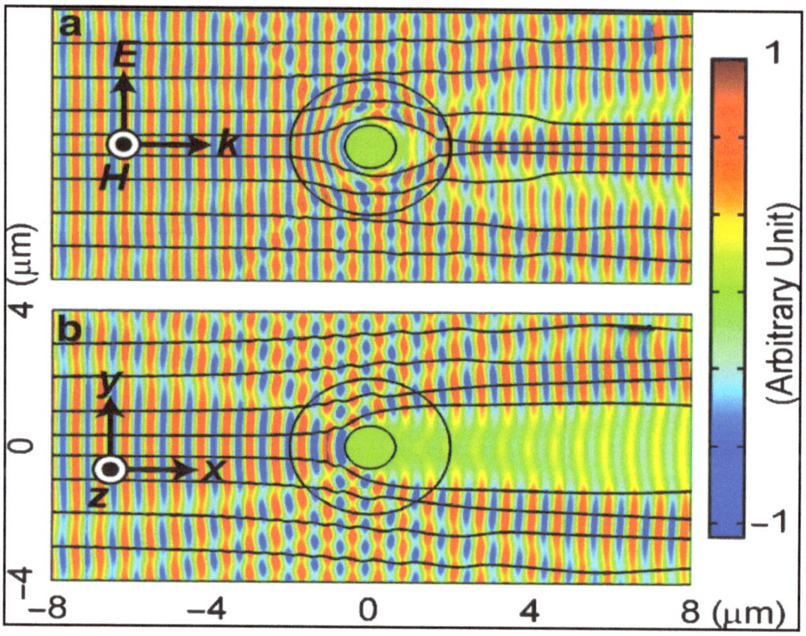

Cloaking Device

11.1.2 Early Experimental Demonstrations

The first experimental demonstrations of cloaking devices primarily occurred in the microwave regime, where fabrication of metamaterials with subwavelength unit cells is relatively straightforward. In 2006 and 2007, pioneering studies used arrays of split-ring resonators and wire structures to create a cylindrical cloak that redirected microwave radiation around an object. These experiments successfully demonstrated that electromagnetic waves could flow around the cloaked region, producing a measurable reduction in scattering. These early demonstrations validated theoretical models and opened a pathway toward more advanced designs, showing that metamaterial-based cloaking was achievable with engineered anisotropic and inhomogeneous materials.

11.1.3 Optical and Infrared Cloaking

Scaling cloaking devices to optical frequencies introduces significant challenges due to the nanometer-scale precision required for the unit cells. Despite these difficulties, researchers have developed metamaterial-based cloaks for near-infrared and visible light using plasmonic and dielectric nanostructures. Optical cloaks often employ transformation optics principles combined with advanced nanofabrication techniques such as electron-beam lithography and focused ion beam milling. Experimental demonstrations have shown partial invisibility, where scattering is reduced and light is guided around micro- and nanoscale objects. While complete invisibility across the entire visible spectrum remains challenging due to material losses and bandwidth limitations, these optical demonstrations represent critical milestones toward practical applications in imaging, photonics, and camouflage.

11.1.4 Acoustic Cloaking Experiments

Cloaking is not limited to electromagnetic waves; acoustic cloaks have also been experimentally demonstrated. These devices manipulate sound waves using metamaterials with spatially varying density and bulk modulus. Initial experiments at audible and ultrasonic frequencies used structured shells composed of subwavelength resonators, labyrinthine channels, or periodic lattices to guide sound around a target. Measurements showed reduced acoustic scattering and sound shadowing behind the cloaked object, confirming the theoretical predictions. Acoustic cloaking holds potential for applications in sonar evasion, noise reduction, and architectural acoustics, offering a complementary avenue to electromagnetic cloaking.

11.1.5 Thermal and Mechanical Cloaking

Thermal cloaking has been experimentally realized by controlling heat conduction rather than wave propagation. By designing materials with anisotropic and spatially graded thermal conductivity, researchers have demonstrated the redirection of heat flow around an object, creating a "thermal shadow" and protecting the cloaked region from temperature changes.

Mechanical cloaking extends similar principles to elastic waves, allowing structures to hide objects from stress or vibration. Experimental setups using elastomeric lattices or layered composites have shown the redirection of mechanical waves, reducing stress concentrations and mitigating vibration effects. These demonstrations highlight the versatility of cloaking concepts beyond traditional electromagnetic applications.

11.1.6 Challenges in Experimental Cloaking

Despite remarkable progress, experimental cloaking remains constrained by several limitations. Bandwidth is a major challenge, as many cloaks operate effectively only over narrow frequency ranges due to resonant structures. Losses in materials, particularly metals and dielectrics, reduce the efficiency of cloaking, especially at optical frequencies.

Fabrication precision is critical, as small deviations can produce scattering and reduce performance. Additionally, most experimental cloaks have been demonstrated for specific polarization, incident angles, or wave types, limiting practical applicability. Scaling to larger objects or three-dimensional, omnidirectional cloaks adds further complexity, requiring advanced design, fabrication, and material solutions.

11.1.7 Recent Advances and Hybrid Approaches

Recent research has explored hybrid approaches to overcome these limitations. Non-resonant cloaks, transformation-free designs, and quasi-conformal mapping techniques reduce losses and expand operational bandwidth. Active cloaks that incorporate sensors and actuators can dynamically adjust their properties to compensate for environmental changes or material imperfections. Advances in 3D printing, nanoscale lithography, and soft materials allow fabri-cation of more complex and flexible cloaking geometries. Additionally, combining electromagnetic, acoustic, and thermal cloaking principles in a single metamaterial structure is an emerging area, enabling multifunctional cloaks capable of manipulating multiple types of waves simultaneously.

11.1.8 Applications of Experimental Cloaking

Experimental cloaking demonstrations provide a foundation for a wide range of applications. In defense and security, cloaking devices could reduce the visibility of vehicles, aircraft, or infrastructure to radar, sonar, or infrared detection systems. In biomedical imaging, cloaks can minimize scattering around implants or sensors, improving image quality and diagnostic accuracy. Acoustic cloaking can enhance architectural acoustics, reduce noise pollution, or improve sonar-based navigation. Thermal cloaking can protect sensitive electronics or mechanical components from localized heating, enhancing reliability and lifespan. While practical deployment remains in early stages, these demonstrations illustrate the potential for transformative applications across multiple domains.

11.2 METASURFACE LENSES IN MOBILE DEVICES

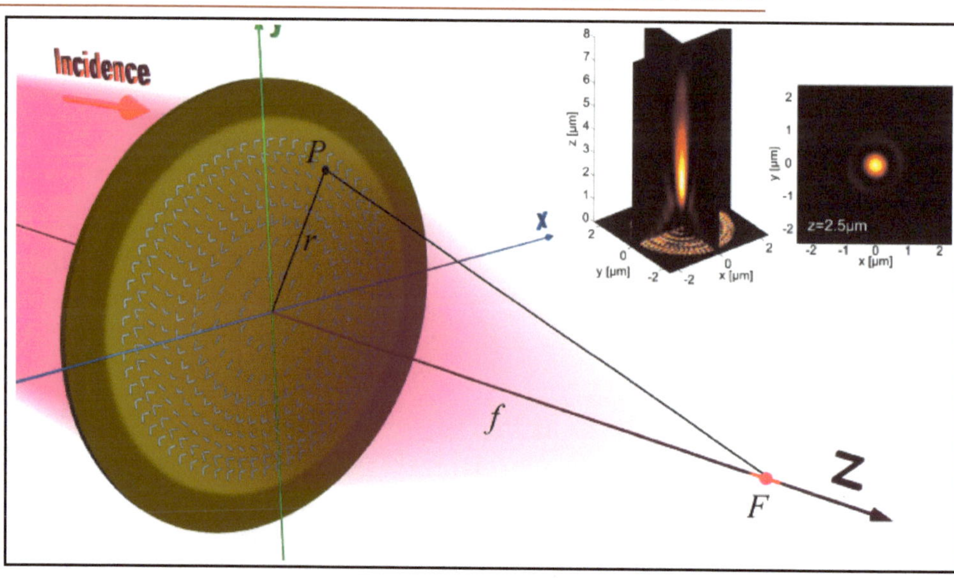

Fig: Metasurface Creates Ultrathin Tiny Planar Lenses

Metasurface lenses represent a transformative advancement in optical engineering, offering unprecedented opportunities for miniaturization, performance enhancement, and multifunctional integration in mobile devices. By manipulating light at subwavelength scales, these flat optics can replace traditional multi-element lenses, reducing thickness and weight while enhancing imaging capabilities. Challenges remain in fabrication, material selection, and broadband efficiency, but ongoing research and industry efforts are steadily overcoming these obstacles. As metasurface lenses mature, they are poised to redefine mobile device optics, enabling compact, high-performance cameras and optical systems that support imaging, augmented reality, and computational photography in ways previously unattainable. The continued integration of metasurface technology promises a new era of innovation in mobile devices, combining sleek form factors with advanced optical functionality.

11.2.1 Principles of Metasurface Lenses

Metasurface lenses rely on arrays of subwavelength scatterers or "meta-atoms" that locally alter the phase, amplitude, or polarization of incident light. Unlike conventional curved lenses that rely on geometric refraction, metasurface lenses implement optical functions through abrupt phase changes across the surface.

By engineering the spatial distribution of these meta-atoms, light can be focused, collimated, or shaped to form arbitrary wavefronts. This approach enables flat, ultrathin lenses capable of high numerical apertures, reduced chromatic aberration, and complex imaging functionalities. The compact form factor of metasurface lenses makes them particularly suited for mobile devices, where space is limited and optical performance is critical.

11.2.2 Advantages for Mobile Devices

The most significant advantage of metasurface lenses in mobile devices is their ultrathin form factor. Traditional camera lenses require multiple curved elements stacked together to correct optical aberrations, leading to protruding camera modules. Metasurface lenses can achieve similar or superior performance within a single, planar layer, dramatically reducing device thickness.

In addition, metasurfaces allow precise correction of optical aberrations, such as chromatic and spherical aberrations, using phase-engineered meta-atoms, reducing the need for complex multi-element assemblies. The lightweight nature of metasurface lenses also contributes to device portability and structural design flexibility. Furthermore, metasurfaces can be engineered to manipulate polarization and phase, enabling functionalities such as augmented reality (AR) displays, depth sensing, and computational imaging within the same optical layer.

11.2.3 Fabrication and Integration Challenges

Fabricating metasurface lenses for mobile devices requires nanometer-scale precision, as the performance depends critically on the geometry, arrangement, and material properties of the meta-atoms. Techniques such as electron-beam lithography, nanoimprinting, and photolithography have been employed to create metasurfaces with high fidelity. However, scaling these methods for mass production while maintaining cost-effectiveness remains a challenge.

Integration into mobile devices also demands compatibility with existing optical stacks, sensors, and packaging processes. Materials used in metasurfaces must balance optical performance with durability, thermal stability, and environmental resistance, ensuring reliable operation over the lifespan of the device. Despite these challenges, ongoing advances in fabrication techniques and material science are steadily bridging the gap between laboratory prototypes and commercial-scale production.

11.2.4 Optical Performance Enhancement

Metasurface lenses enhance mobile imaging performance in several key ways. By precisely controlling the phase profile of light, they can achieve high numerical apertures, improving light collection efficiency and low-light imaging capabilities. Flat optics can also reduce chromatic aberration and spherical distortions, delivering sharper images across the field of view. Additionally, metasurfaces enable complex wavefront engineering, such as beam shaping and polarization manipulation, which can enhance features like depth sensing, 3D imaging, and computational photography. These capabilities allow mobile devices to perform tasks that traditionally required bulky optical systems, such as high-resolution zooming, portrait mode imaging, and AR integration, without compromising form factor.

11.2.5 Multifunctional Optical Components

One of the most transformative aspects of metasurface lenses is their ability to serve multiple optical functions simultaneously. Traditional lenses often require separate elements to focus, filter, or manipulate light, increasing complexity and size. Metasurfaces can integrate multiple functionalities into a single layer, combining focusing, polarization control, spectral filtering, and beam shaping. In mobile devices, this integration reduces the number of optical elements, enhances system efficiency, and supports advanced computational imaging techniques. For instance, a metasurface lens can simultaneously focus light onto the sensor while filtering specific wavelengths for enhanced color reproduction or integrating diffractive elements for augmented reality displays. This multifunctionality is key to enabling the next generation of compact, high-performance mobile optics.

11.2.6 Impact on Device Design and Miniaturization

The adoption of metasurface lenses has profound implications for mobile device design. By replacing traditional multi-element lenses with ultrathin, planar optics, device manufacturers can achieve slimmer camera modules, reducing the camera bump that has become a common design concern.

This miniaturization allows for sleeker, lighter smartphones while maintaining or even improving imaging capabilities. The reduced weight and volume also create opportunities for integrating additional sensors, batteries, or other components, enhancing overall device functionality. In wearable devices, such as smart glasses or AR headsets, metasurface lenses are particularly valuable, enabling compact optical systems that would otherwise be impractical.

11.2.7 Challenges and Future Directions

Despite their potential, several challenges remain before metasurface lenses become ubiquitous in mobile devices. Mass fabrication at low cost, scalability to large areas, and maintaining high optical efficiency across the visible spectrum are ongoing challenges. Material limitations, particularly concerning absorption, environmental stability, and mechanical robustness, must be addressed.

Hybrid approaches, combining metasurfaces with conventional optics or active tunable components, are being explored to enhance performance and adaptability. Future research is also focused on broadband operation, dynamic focusing, and integration with computational imaging algorithms, enabling even more advanced imaging functionalities within the constrained space of mobile devices.

11.2.8 Commercialization and Industry Adoption

Several companies and research groups have already demonstrated prototype metasurface lenses for mobile devices, highlighting their commercial potential. As fabrication techniques mature and cost-effectiveness improves, these lenses are expected to be integrated into mainstream smartphones and wearable devices. Beyond imaging, metasurface lenses may enable new applications such as miniaturized LiDAR systems, augmented reality displays, and advanced optical sensors, driving innovation across consumer electronics. The combination of high performance, compact form factor, and multifunctionality positions metasurface lenses as a critical technology for the next generation of mobile devices.

11.3 ACOUSTIC METAMATERIALS IN ARCHITECTURE

Acoustic metamaterials offer a revolutionary approach to architectural acoustics, enabling precise control over sound propagation, vibration, and noise management. By leveraging engineered subwavelength structures, these materials provide solutions that are thinner, more efficient, and more versatile than conventional methods. Applications range from noise reduction in urban environments to optimized sound distribution in concert halls, theaters, and office spaces.

While challenges in fabrication, scalability, and integration remain, ongoing research and experimental demonstrations highlight the transformative potential of acoustic metamaterials in architecture. As fabrication techniques advance and interdisciplinary collaboration continues, acoustic metamaterials are poised to become an integral part of modern architectural design, combining functionality, aesthetics, and adaptability to create spaces that are both acoustically optimized and visually inspiring.

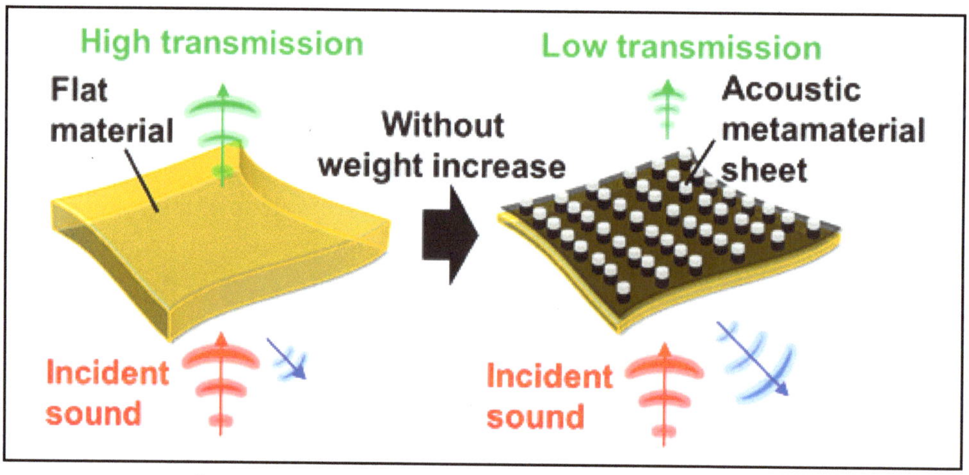

Acoustic Metamaterials Based on Polymer Sheets

11.3.1 Principles of Acoustic Metamaterials

Acoustic metamaterials are engineered structures designed to influence sound waves through their geometry and internal structure rather than their chemical composition alone. By arranging subwavelength resonators, channels, or lattice structures in specific patterns, these materials can create effective acoustic properties such as negative mass density or negative bulk modulus. These properties enable phenomena like sound redirection, sound cloaking, and selective frequency filtering. The fundamental principle is that the periodic or resonant structures interact with incident sound waves, producing interference patterns and energy localization that conventional materials cannot achieve. The versatility of acoustic metamaterials allows architects and engineers to tailor the acoustic response of spaces for specific needs, from concert halls to office buildings.

11.3.2 Noise Reduction and Sound Isolation

One of the primary applications of acoustic metamaterials in architecture is noise reduction. Traditional methods, such as thick walls, insulation, and absorptive panels, often require significant material and space, limiting design flexibility.

Acoustic metamaterials, on the other hand, can achieve high-performance sound attenuation in much thinner structures by selectively blocking or redirecting specific frequencies. For instance, metamaterials with tunable band gaps can prevent unwanted sound from penetrating walls or ceilings while allowing other frequencies to pass, creating acoustically optimized environments. This capability is especially valuable in urban architecture, where traffic, construction, and industrial noise can degrade indoor comfort. By integrating acoustic metamaterials into building envelopes, architects can achieve superior sound isolation without compromising aesthetic design or spatial efficiency.

11.3.3 Enhancing Room Acoustics

Beyond noise reduction, acoustic metamaterials can enhance room acoustics by shaping sound propagation within interior spaces. In concert halls, theaters, and lecture rooms, controlling reverberation, echo, and frequency distribution is critical for auditory clarity and sound quality. Metamaterials can be engineered to focus sound, distribute it evenly, or suppress unwanted resonances, creating tailored acoustic environments. For example, metasurfaces with subwavelength corrugations or resonators can redirect sound waves to minimize dead zones and enhance the listening experience. This level of control allows architects to design spaces that not only meet functional acoustic requirements but also improve the emotional and aesthetic experience of occupants, enhancing the overall architectural quality.

11.3.4 Aesthetic Integration and Architectural Design

Acoustic metamaterials provide architects with opportunities to integrate functional acoustic elements seamlessly into building design. Unlike traditional bulky panels or insulation layers, metamaterials can be designed as thin, visually appealing surfaces that also serve acoustic purposes. Patterns, textures, and geometries of acoustic metamaterials can be customized to complement interior design, allowing spaces to be both acoustically optimized and visually striking. This integration supports innovative architectural concepts, where sound management is an inherent part of the building's aesthetics rather than an add-on feature. The ability to combine form and function aligns with contemporary architectural trends emphasizing efficiency, elegance, and user experience.

11.3.5 Vibration Isolation and Structural Acoustics

Acoustic metamaterials also offer solutions for structural vibration control, an important aspect of architectural acoustics. Buildings, bridges, and floors can transmit vibrations caused by mechanical systems, footsteps, or environmental forces, which can compromise comfort and structural integrity. Metamaterials with negative effective mass or tunable resonance can absorb or redirect these vibrations, effectively isolating sensitive areas from disruptive mechanical energy. Incorporating these materials into floors, walls, or ceilings can enhance structural acoustics, improving both human comfort and the durability of architectural elements. This capability is particularly valuable in high-rise buildings, hospitals, and laboratories, where precision and quiet operation are essential.

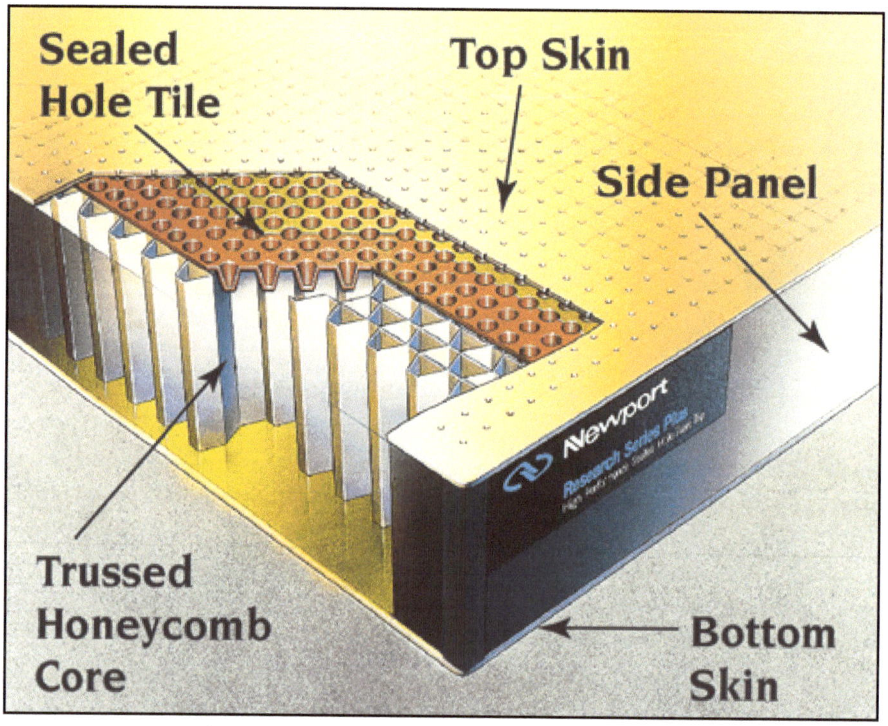

Vibration Isolated Rigid Support Structures

11.3.6 Challenges in Implementation

Despite their advantages, implementing acoustic metamaterials in architecture faces several challenges. Fabrication at scale remains a major hurdle, as precise subwavelength structures are required to achieve desired acoustic effects. Material durability, cost, and environmental stability must also be considered, particularly for large-scale building applications. Additionally, designing metamaterials for broadband sound control and multi-directional sound propagation requires advanced computational modeling and acoustic simulations. Integrating these materials into existing construction methods and ensuring compliance with building codes and regulations adds further complexity. Overcoming these challenges requires collaboration between architects, acoustic engineers, material scientists, and construction specialists.

11.3.7 Recent Experimental Demonstrations

Recent studies have demonstrated the effectiveness of acoustic metamaterials in architectural applications. Experimental installations in office spaces, concert halls, and urban sound barriers have shown significant reductions in noise levels, improved sound clarity, and enhanced vibration isolation. For example, labyrinthine acoustic metamaterials have been used to block low-frequency urban noise in compact barriers, while metasurface panels in auditoriums have successfully managed reflections and reverberations to optimize auditory experiences. These experiments highlight the practical potential of acoustic metamaterials in real-world architectural environments and provide benchmarks for future large-scale implementations.

11.3.8 Future Directions

The future of acoustic metamaterials in architecture is closely linked to advancements in fabrication, tunability, and multifunctionality. Development of 3D-printed or modular metamaterials can enable cost-effective large-scale production, while active metamaterials with tunable properties can adapt to changing acoustic environments. Integration with smart building systems, sensors, and IoT technology will allow real-time acoustic optimization, dynamically controlling noise, reverberation, and vibration based on occupancy and activity. Bio-inspired designs, mimicking natural structures such as bird feathers or termite mounds, can provide efficient sound management with minimal material usage. These future directions promise not only enhanced acoustic performance but also sustainable, energy-efficient, and adaptive architectural solutions.

11.4 MEDICAL DIAGNOSTICS AND IMAGING ENHANCEMENTS

Metamaterials offer transformative opportunities for medical diagnostics and imaging, enabling subwavelength resolution, enhanced signal contrast, and multifunctional imaging capabilities. By controlling electromagnetic, acoustic, and optical waves in ways unattainable with natural materials, metamaterials improve diagnostic accuracy, reduce device size, and enable portable and wearable imaging solutions. Challenges remain in fabrication, biocompatibility, and regulatory approval, but experimental demonstrations and ongoing research illustrate significant potential for clinical translation. As metamaterials continue to evolve, they promise to revolutionize medical imaging, supporting earlier disease detection, more accurate diagnostics, and personalized healthcare solutions in a wide range of clinical contexts.

11.4.1 Principles of Metamaterial-Enhanced Imaging

Metamaterials achieve enhanced imaging through their ability to manipulate wave propagation beyond the limits of natural materials. Electromagnetic metamaterials can be engineered to achieve negative refractive indices, superlensing, and near-field focusing, overcoming diffraction limits and enabling subwavelength imaging.

Acoustic metamaterials allow control over sound waves, improving ultrasound imaging resolution and enabling the redirection or concentration of acoustic energy. Optical metamaterials can manipulate light at nanoscales, enhancing techniques such as optical coherence tomography (OCT) and endoscopic imaging. The fundamental principle is that engineered structures, often composed of subwavelength resonators or lattices, interact with incident waves to focus, filter, or amplify signals in ways that traditional lenses or waveguides cannot achieve.

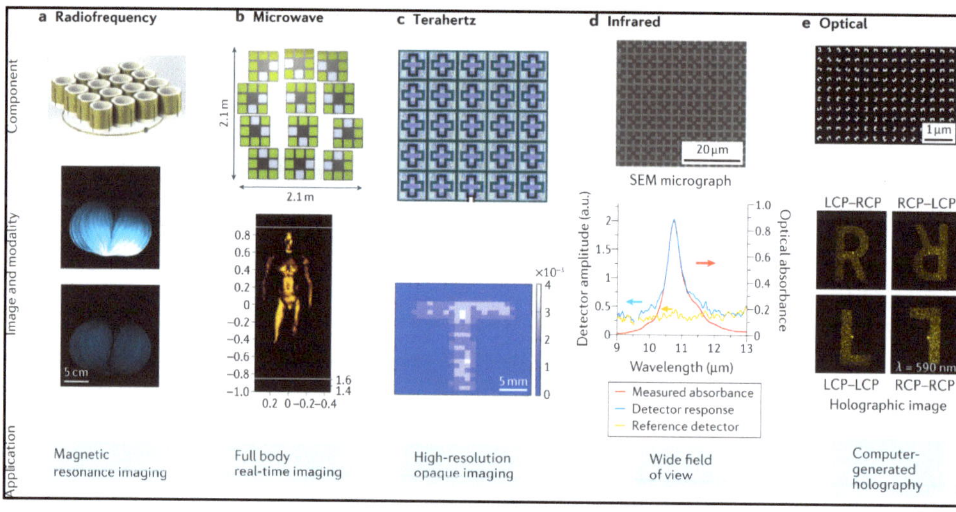

Imaging with Metamaterials

11.4.2 Superlensing for Subwavelength Imaging

One of the most significant contributions of metamaterials to medical imaging is superlensing, which allows imaging below the diffraction limit. Conventional lenses are constrained by diffraction, limiting resolution to approximately half the wavelength of the probing wave.

Metamaterial-based superlenses, often constructed with negative-index materials or plasmonic structures, can focus electromagnetic waves beyond this limit, enabling detailed visualization of cellular structures or microvascular networks. In optical imaging, superlenses can enhance endoscopy or OCT, providing higher resolution images that allow earlier detection of abnormalities such as tumors, lesions, or vascular irregularities. In MRI and X-ray imaging, metamaterial-inspired designs can improve signal concentration, enhancing image clarity without increasing exposure or scan duration.

11.4.3 Enhanced Ultrasound Imaging

Acoustic metamaterials have a transformative impact on ultrasound imaging, which is widely used for soft tissue examination, cardiac monitoring, and prenatal imaging. By incorporating resonant structures, waveguides, or gradient-density materials, acoustic metamaterials can focus sound waves with high precision, reduce scattering, and improve penetration depth. These enhancements increase image resolution and contrast, enabling better visualization of small anatomical features such as blood vessels, nerve fibers, or microcalcifications. Tunable acoustic metamaterials also allow dynamic control over focal points, supporting real-time adjustments during imaging procedures, which is critical for diagnostic accuracy and patient safety.

11.4.4 Contrast Enhancement and Signal Amplification

Metamaterials can also amplify or manipulate signals to enhance contrast in medical imaging. In optical imaging, plasmonic metamaterials concentrate electromagnetic fields at subwavelength scales, boosting the sensitivity of techniques like fluorescence microscopy and Raman spectroscopy. This allows detection of low-abundance biomarkers or early-stage pathological changes. In MRI, metamaterials designed to focus radiofrequency fields can increase signal-to-noise ratio, improving the visibility of fine structures without increasing magnetic field strength. By selectively enhancing desired signals and suppressing background noise, metamaterials improve diagnostic accuracy, reduce false positives, and facilitate early intervention.

11.4.5 Miniaturization and Portable Imaging Devices

Another critical advantage of metamaterials in medical diagnostics is the potential for miniaturization. Conventional imaging systems often require bulky lenses, waveguides, or transducers, limiting portability and accessibility. Metamaterial-based lenses and waveguides can achieve high performance in ultrathin, flat, or flexible formats, enabling compact imaging devices suitable for point-of-care diagnostics, wearable health monitors, or handheld ultrasound units. This miniaturization facilitates rapid diagnostics in remote or resource-limited settings, improving healthcare delivery and patient outcomes. Flexible metamaterials can conform to the human body, allowing closer contact and more accurate signal capture in applications such as skin imaging or cardiovascular monitoring.

11.4.6 Multimodal Imaging and Functional Integration

Metamaterials also support multimodal imaging, where multiple diagnostic techniques are combined to provide comprehensive clinical information. For example, hybrid optical-acoustic metamaterials can integrate ultrasound and optical imaging in a single device, allowing simultaneous structural and functional assessment. Electromagnetic metamaterials can enable concurrent MRI and radiofrequency-based spectroscopy, providing anatomical and molecular information in one scan. Such integration reduces the need for multiple procedures, improves diagnostic efficiency, and enables richer data collection for personalized medicine. The design flexibility of metamaterials allows the creation of multifunctional imaging surfaces that can be tuned to specific diagnostic requirements.

11.4.7 Challenges in Clinical Implementation

Despite their promise, the clinical adoption of metamaterials in medical imaging faces several challenges. Fabrication precision is critical, as subwavelength structures require nanoscale accuracy to achieve desired properties. Biocompatibility, sterilization, and mechanical stability are essential for safe use in medical environments. Additionally, ensuring broadband functionality and minimal losses is important for imaging across different modalities and patient conditions. Regulatory approval, integration with existing medical equipment, and cost-effectiveness also influence the feasibility of clinical translation. Ongoing research focuses on developing scalable fabrication techniques, durable materials, and hybrid designs that meet clinical standards while maintaining performance.

11.4.8 Recent Experimental Demonstrations

Experimental studies have demonstrated the effectiveness of metamaterials in enhancing medical imaging. Plasmonic superlenses have resolved features below 100 nanometers in optical microscopy. Acoustic metamaterial lenses have improved ultrasound resolution in soft tissue phantoms and in vivo animal models. Radiofrequency metamaterials have increased MRI signal-to-noise ratio and reduced imaging artifacts. These demonstrations validate the theoretical potential of metamaterials and provide a foundation for further development toward human clinical applications. Additionally, prototype devices incorporating flexible or wearable metamaterials show promise for real-time, non-invasive diagnostics in dynamic environments.

11.4.9 Future Directions

The future of metamaterials in medical diagnostics and imaging is oriented toward fully integrated, adaptive, and intelligent systems. Research is exploring tunable and active metamaterials that can adjust their optical or acoustic properties in real time, enhancing imaging in variable physiological conditions. 4D metamaterials, which change their structure dynamically, may enable adaptive focusing, contrast enhancement, or wave steering.

Bio-inspired designs, mimicking natural structures such as photoreceptors or cochlear mechanics, offer opportunities for more efficient, sensitive, and multifunctional imaging systems. Combining metamaterials with computational imaging, artificial intelligence, and machine learning will further optimize image acquisition, reconstruction, and interpretation, creating a new paradigm in precision diagnostics.

11.5 SMART INFRASTRUCTURE AND VIBRATION MANAGEMENT

- **Principles of Vibration Control with Metamaterials:** Vibration management using metamaterials relies on the ability to control the propagation of elastic and mechanical waves through engineered structures. Mechanical metamaterials can exhibit properties such as negative effective mass, negative stiffness, or tunable band gaps, which allow them to block, redirect, or attenuate vibrations at specific frequencies. Phononic crystals, a type of mechanical metamaterial, create periodic structures that generate band gaps, preventing the propagation of vibrations within certain frequency ranges. By integrating these materials into buildings, bridges, or mechanical systems, engineers can selectively isolate critical components from disruptive vibrations, enhancing stability and performance without adding significant mass or bulk.
- **Applications in Structural Vibration Isolation:** One of the primary applications of metamaterials in smart infrastructure is vibration isolation. Buildings and bridges experience vibrations from environmental factors such as wind, earthquakes, and traffic. Conventional isolation systems rely on passive damping or tuned mass dampers, which are effective only over limited frequency ranges. Metamaterials, with their tunable band gaps and frequency-selective attenuation, offer a more versatile solution. For example, a phononic crystal floor or wall can prevent specific vibrational frequencies from propagating through a building, reducing noise and mechanical stress on structural components. This approach enhances occupant comfort and extends the lifespan of infrastructure by mitigating fatigue and resonance-related damage.
- **Adaptive and Tunable Vibration Management:** Beyond static vibration control, metamaterials enable adaptive and tunable responses. Smart mechanical metamaterials can incorporate sensors, actuators, or stimuli-responsive elements that adjust their properties in real time based on external inputs. For instance, materials that change stiffness or resonance characteristics in response to applied stress, temperature, or electromagnetic fields can dynamically alter vibration transmission pathways. Such adaptability is crucial in infrastructure subjected to variable loads, such as bridges experiencing fluctuating traffic or buildings exposed to shifting wind conditions. By actively tuning vibration responses, smart metamaterials provide real-time mitigation, improving safety and performance under diverse operating conditions.

- **Integration into Civil Engineering Structures:** The integration of metamaterials into civil infrastructure requires careful consideration of design, material compatibility, and construction methods. Structural elements such as beams, columns, and floors can be enhanced with embedded phononic crystals or metamaterial layers to attenuate specific vibration modes. Bridge decks, for example, can incorporate metamaterial arrays to reduce traffic-induced vibrations and minimize fatigue on supporting structures. In high-rise buildings, metamaterial-based isolators can be integrated into floors or walls to reduce sway and vibration transmission. These systems offer advantages in terms of compactness, lightweight design, and frequency-specific control, complementing traditional engineering approaches while enabling more responsive and efficient vibration management.
- **Enhancing Seismic Resilience:** Metamaterials also have potential applications in seismic vibration mitigation. Earthquake-resistant design traditionally relies on base isolators, damping systems, and structural reinforcement to absorb and dissipate seismic energy. Mechanical metamaterials, particularly those with tunable band gaps and negative effective mass, can be engineered to redirect seismic waves away from critical structural components or attenuate specific frequency ranges associated with resonant amplification. By integrating metamaterial layers within foundations, walls, or floors, buildings can achieve enhanced seismic resilience, reducing structural damage and improving safety during earthquakes. This approach complements conventional seismic engineering techniques and introduces new strategies for wave manipulation at the structural scale.
- **Vibration Management in Transportation Infrastructure:** Transportation infrastructure, including bridges, railways, and highways, benefits significantly from metamaterial-based vibration management. Vibrations induced by vehicles, trains, and machinery can accelerate material fatigue, reduce structural lifespan, and compromise safety. Acoustic and mechanical metamaterials can be embedded into bridge decks or railway tracks to isolate vibrations and dampen high-amplitude oscillations. Phononic crystal barriers along railways can reduce ground-borne vibrations transmitted to nearby buildings, improving urban living conditions and mitigating environmental impact. These metamaterial applications enhance durability, reduce maintenance costs, and improve the performance of transportation networks in densely populated areas.

- **Challenges in Implementation:** While the potential of metamaterials in vibration management is substantial, practical implementation faces challenges. Fabrication of large-scale metamaterial structures requires precision engineering, particularly for phononic crystals with subwavelength features. Material durability, environmental resistance, and long-term stability are critical considerations for infrastructure applications exposed to weather, temperature fluctuations, and mechanical stress. Integration with conventional construction methods, cost-effectiveness, and compliance with building codes are additional hurdles. Advanced computational modeling, multidisciplinary design approaches, and scalable manufacturing techniques are essential to translate laboratory-scale demonstrations into real-world infrastructure solutions.
- **Experimental Demonstrations and Case Studies:** Experimental studies have demonstrated the effectiveness of metamaterials in vibration control for civil and mechanical structures. Laboratory-scale phononic crystal plates and beams have successfully attenuated vibrations over targeted frequency ranges. Full-scale bridge models incorporating metamaterial isolators have shown reduced oscillations under simulated traffic loads. Smart adaptive metamaterials with tunable stiffness have demonstrated real-time control over vibrational propagation, confirming theoretical predictions. These experimental successes provide proof-of-concept validation and guide the development of scalable, practical solutions for smart infrastructure applications.

CHAPTER 12

Appendices

12.1 GLOSSARY OF KEY TERMS

Absorption Coefficient: A measure of how much incident wave energy (electromagnetic, acoustic, or thermal) is absorbed by a material rather than transmitted or reflected.

Acoustic Metamaterials: Engineered materials designed to control, direct, and manipulate sound waves in ways not achievable with conventional materials, often exhibiting negative mass density or bulk modulus.

Anisotropy: A property of a material in which its characteristics, such as permittivity, permeability, or thermal conductivity, vary depending on the direction of measurement.

Band Gap: A range of frequencies in which wave propagation (electromagnetic or acoustic) is forbidden due to the periodic structure of a material.

Bio-Inspired Metamaterials: Metamaterials designed based on natural structures or biological principles to achieve optimized wave control, robustness, or adaptability.

Bulk Modulus: A measure of a material's resistance to uniform compression; in metamaterials, it can be engineered to negative values to control wave propagation.

Chirality: A property of a structure that is not superimposable on its mirror image, leading to unique optical effects such as circular dichroism in metamaterials.

Cloaking: The process of rendering an object invisible to waves (electromagnetic, acoustic, or thermal) by guiding the waves around it using metamaterials.

Composite Material: A material composed of two or more constituent materials with distinct properties, often used in metamaterials to achieve desired effective behavior.

Computational Electromagnetics: The use of numerical methods to model and analyze the interaction of electromagnetic waves with metamaterials.

Dielectric Metamaterials: Metamaterials constructed from non-conducting materials that manipulate electromagnetic waves through engineered permittivity rather than metallic resonance.

Effective Medium Theory (EMT): A theoretical approach to approximate the macroscopic properties of a heterogeneous metamaterial by treating it as a uniform medium.

Electromagnetic Metamaterials: Engineered materials that manipulate electromagnetic waves, often exhibiting unusual properties such as negative refractive index or epsilon-near-zero behavior.

Epsilon-Near-Zero (ENZ): A condition in which the permittivity of a material approaches zero, allowing unique wave propagation effects such as phase uniformity and tunneling.

Fabry-Pérot Resonance: A resonance phenomenon occurring when waves reflect multiple times between parallel surfaces, often used in metamaterial designs to enhance wave interactions.

Flat Optics: Optical devices, such as metasurfaces, that control light using planar, ultrathin structures instead of traditional curved lenses.

Frequency Selective Surface (FSS): A surface with periodic structures that selectively transmit or reflect electromagnetic waves at specific frequencies.

Gradient Index (GRIN): A refractive index profile that gradually varies across a material, used in metamaterials to bend or focus waves.

Hyperbolic Metamaterials: Materials with extreme anisotropy where permittivity or permeability has opposite signs along different axes, enabling subwavelength imaging and enhanced light-matter interaction.

Left-Handed Material (LHM): A metamaterial in which the electric field, magnetic field, and wave vector form a left-handed system, often exhibiting negative refractive index.

Local Resonance: Phenomenon where subwavelength structures resonate with incoming waves, creating band gaps and anomalous wave propagation in metamaterials.

Magnetic Metamaterials: Metamaterials designed to control magnetic fields or exhibit engineered permeability, often using split-ring resonators (SRRs).

Metamaterial Lens (Metasurface Lens): A lens constructed from metasurfaces capable of focusing or shaping light with subwavelength precision while maintaining a flat, ultrathin profile.

Metasurface: A two-dimensional array of engineered structures that manipulate wavefronts, phase, amplitude, or polarization of incident waves.

Near-Field Imaging: Imaging technique that captures evanescent waves close to an object, often enhanced by metamaterials to achieve subwavelength resolution.

Negative Refraction: A phenomenon where waves refract in the opposite direction compared to conventional materials, enabling superlensing and cloaking applications.

Phononic Crystal: A periodic material designed to control acoustic or elastic waves, creating band gaps and manipulating vibrations.

Photonic Crystal: A periodic optical nanostructure that affects the motion of photons, creating band gaps for light similar to electronic band gaps in semiconductors.

Plasmonics: The study and application of surface plasmons, collective oscillations of electrons at metal-dielectric interfaces, often used in optical metamaterials.

Polarization Control: The manipulation of the orientation of waves' electric or magnetic fields using metamaterials for applications such as lenses, filters, or sensors.

Programmable Metamaterials: Metamaterials whose properties can be dynamically altered through external stimuli such as voltage, magnetic fields, or mechanical deformation.

Resonant Frequency: The specific frequency at which a metamaterial structure naturally oscillates, often critical for wave control and band gap creation.

Split-Ring Resonator (SRR): A commonly used metamaterial unit cell that creates strong magnetic resonance at subwavelength scales, enabling negative permeability and other unusual electromagnetic properties.

Subwavelength Structure: A feature or element smaller than the wavelength of the wave interacting with it, essential for achieving metamaterial properties.

Superlensing: A lensing effect enabled by metamaterials that overcomes the diffraction limit, allowing imaging with resolution finer than the wavelength of light.

Tunable Metamaterials: Metamaterials that allow controlled modification of their properties, enabling adaptive control over wave propagation for applications such as sensing, imaging, and cloaking.

Wavefront Shaping: The process of altering the phase, amplitude, or direction of waves using engineered structures to achieve desired focusing, redirection, or interference effects.

Wireless Signal Manipulation: The application of metamaterials to control, enhance, or filter electromagnetic waves in communication systems, improving efficiency and reducing interference.

Zero-Index Metamaterials: Materials in which the refractive index approaches zero, allowing unusual wave propagation phenomena like infinite phase velocity or wave tunneling.

12.2 LIST OF KEY JOURNALS AND CONFERENCES

12.2.1 Key Academic Journals in Metamaterials

High-Impact, Broad Scope Journals

- **Nature:** Frequently publishes groundbreaking, high-impact research and reviews in photonics and metamaterials.
- **Science:** Features seminal papers on transformative metamaterial concepts, particularly in electromagnetic and optical metamaterials.
- **Nature Materials:** Covers fundamental advances in materials science, including the development and characterization of new metamaterials.
- **Nature Photonics:** A premier journal for research in photonics, regularly featuring articles on optical and plasmonic metamaterials, metasurfaces, and transformation optics.
- **Nature Physics:** Publies influential research on the fundamental physical principles underlying metamaterial phenomena.
- **Physical Review Letters (PRL):** A top-tier journal for publishing short, high-impact results across physics, including many foundational papers on metamaterial theory and experimental demonstrations.
- **Advanced Materials:** A leading interdisciplinary journal covering all aspects of materials science, with a strong focus on novel functional materials, including metamaterials.
- **ACS Nano:** Publishes comprehensive research on nanoscience and nanotechnology, encompassing the nanofabrication and characterization of metamaterials.
- **Nano Letters:** Features cutting-edge research in nanoscience, with a significant number of publications on plasmonics and metamaterial devices.
- **Light: Science & Applications (Nature Partner Journal):** Dedicated to publishing top-tier research in all areas of optics and photonics, with a strong emphasis on metamaterials and nanophotonics.

Specialized and Field-Specific Journals

- **Optica (formerly OSA):** The flagship journal of Optica (formerly OSA), publishing high-impact results across all fields of optics and photonics, including metamaterials.

- **Laser & Photonics Reviews:** Publishes review articles, research articles, and perspectives in photonics and optics, with extensive coverage of metamaterials.
- **Advanced Optical Materials:** An interdisciplinary forum for peer-reviewed research on light-matter interactions, a natural home for metamaterials and metasurfaces research.
- **Physical Review Applied:** Focuses on high-quality research at the intersection of physics and engineering, publishing many papers on applied metamaterial devices.
- **Physical Review B:** Condensed Matter and Materials Physics: The primary journal for theoretical and experimental research on the physics of condensed matter, including the electronic and photonic properties of metamaterials.
- **IEEE Transactions on Antennas and Propagation:** A key journal for research on antenna theory, design, and applications, heavily featuring metamaterial-based antennas (metamaterial lenses, leaky-wave antennas, etc.).
- **IEEE Transactions on Microwave Theory and Techniques:** The leading forum for sharing research in microwave engineering, including metamaterials for microwave applications, filters, and waveguides.
- **Applied Physics Letters (APL):** Rapid publication of concise, significant new findings in applied physics, a common venue for new metamaterial concepts and experimental results.
- **Journal of Applied Physics:** Publishes original research on applied physics, covering the broad experimental and theoretical aspects of metamaterials.
- **New Journal of Physics:** An open-access journal publishing across all areas of physics, with a strong section on photonics and metamaterials.
- **Nanophotonics:** Focuses on the interaction of photons with nano-structures, including metamaterials, plasmonics, and silicon photonics.
- **Materials Today:** Publishes authoritative reviews and original research on materials science, often covering progress in metamaterials.
- **APL Materials (AIP Publishing):** An open-access journal that highlights the intersection of structure, properties, and functionality in materials, including metamaterials.

- **Advanced Functional Materials:** Publishes research on materials science that advances the understanding of function, a key theme in metamaterials research.
- **Scientific Reports (Nature Portfolio):** A multidisciplinary open-access journal that publishes a large volume of solid research in metamaterials and nanophotonics.
- **IEEE Access:** A multidisciplinary open-access journal that features a significant number of applied engineering studies on metamaterials.

12.2.2 Key Conferences in Metamaterials

Major Annual Conferences

- **International Congress on Artificial Materials for Novel Wave Phenomena (Metamaterials):** This is the premier annual conference series dedicated exclusively to metamaterials. It covers a broad spectrum of topics from fundamental theory to applications.
- **Conference on Lasers and Electro-Optics (CLEO) / Europe - European Quantum Electronics Conference (EQEC):** A leading global event on lasers and electro-optics, with extensive sessions on metamaterials, metasurfaces, and nanophotonics.
- **SPIE Photonics Europe:** A major biennial conference in Europe featuring numerous tracks on metamaterials, photonic crystals, and plasmonics.
- **SPIE Optics + Photonics:** A large annual event in the United States with several conferences dedicated to nanophotonics, metamaterials, and plasmonics.
- **Materials Research Society (MRS) Spring and Fall Meetings:** While broader in scope, these meetings consistently feature multiple symposia on metamaterials, programmable matter, and architected materials.
- **IEEE International Symposium on Antennas and Propagation and USNC-URSI Radio Science Meeting:** The most significant annual conference for antenna research, featuring a vast amount of work on metamaterial-inspired and metamaterial-based antennas.
- **IEEE MTT-S International Microwave Symposium (IMS):** The world's premier annual conference for microwave engineers, with a strong focus on metamaterials for RF and microwave applications (filters, phase shifters, absorbers).

- **Progress In Electromagnetics Research Symposium (PIERS):** A large international conference on electromagnetic theory and applications, with substantial content on metamaterials.

Other Significant and Biennial Conferences

- **International Conference on Surface Plasmon Photonics (SPP):** A top conference focused specifically on plasmonics, which is deeply intertwined with optical metamaterials research.
- **EOS Annual Meeting (Topical Meeting on Nanophotonics and Metamaterials):** Organized by the European Optical Society (EOS), this meeting is a key European event for the field.
- **International Conference on Phonon Scattering in Condensed Matter (Phonons):** Highly relevant for the acoustic metamaterials community.
- **International Conference on Metamaterials and Photonic Crystals (METAMAT):** A focused conference on the intersection of these two fields.
- **International Conference on Optical Metamaterials (Metamaterials):** Another dedicated conference series that occurs in various locations.

12.3 RESEARCH LABS AND INSTITUTES WORKING ON METAMATERIALS

Institution	Location	Focus Area	Notable Contributions	Key Researchers
Massachusetts Institute of Technology (MIT)	USA	Electromagnetic metamaterials, acoustic metamaterials	Developed first 3D printed acoustic metamaterials, advanced transformation optics	Prof. Steven Johnson, Prof. Marin Soljačić
California Institute of Technology (Caltech)	USA	Optical metamaterials, plasmonics	Pioneered work on hyperbolic metamaterials, plasmonic nanostructures	Prof. Harry Atwater, Prof. Andrei Faraon
University of California, Berkeley	USA	Mechanical metamaterials, thermal metamaterials	Developed ultra-lightweight mechanical metamaterials, thermal cloaking	Prof. Xiang Zhang, Prof. Chris Dames

Harvard University	USA	Bio-inspired metamaterials, reconfigurable metamaterials	Created first large-scale metamaterial structures, adaptive optical systems	Prof. Federico Capasso, Prof. Joanna Aizenberg
Imperial College London	UK	Acoustic metamaterials, photonic crystals	Advanced work on sonic crystals, developed new photonic bandgap materials	Prof. John Pendry, Prof. Stefan Maier
University of Cambridge	UK	Optical metamaterials, quantum metamaterials	Pioneered work on quantum metamaterials, developed new plasmonic devices	Prof. Jeremy Baumberg, Prof. Ulrich Steiner
Karlsruhe Institute of Technology (KIT)	Germany	Electromagnetic metamaterials, terahertz metamaterials	Developed advanced terahertz metamaterials, new electromagnetic wave manipulation techniques	Prof. Martin Wegener, Prof. Carsten Rockstuhl
Max Planck Institute for Intelligent Systems	Germany	Mechanical metamaterials, active metamaterials	Created first mechanical metamaterials with negative Poisson's ratio, developed active tunable systems	Prof. Peer Fischer, Prof. Metin Sitti
University of Pennsylvania	USA	Structural metamaterials, bio-inspired metamaterials	Developed new structural metamaterials, bio-inspired optical systems	Prof. Nader Engheta, Prof. Shu Yang
Purdue University	USA	Thermal metamaterials, multifunctional metamaterials	Pioneered work on thermal cloaking, developed multifunctional metamaterials	Prof. Xiulin Ruan, Prof. Vladimir Shalaev
University of Texas at Austin	USA	Plasmonic metamaterials, nonlinear metamaterials	Advanced work on plasmonic nanostructures, developed nonlinear optical metamaterials	Prof. Andrea Alù, Prof. Mikhail Belkin
ETH Zurich	Switzerland	Acoustic metamaterials, mechanical metamaterials	Developed new acoustic wave manipulation techniques, advanced mechanical systems	Prof. Chiara Daraio, Prof. Dimos Poulikakos
National University of Singapore (NUS)	Singapore	Optical metamaterials, quantum metamaterials	Pioneered work on quantum plasmonics, developed new optical manipulation techniques	Prof. Cheng-Wei Qiu, Prof. Joel Yang

Tsinghua University	China	Electromagnetic metamaterials, terahertz metamaterials	Advanced work on terahertz wave manipulation, developed new electromagnetic systems	Prof. Shining Zhu, Prof. Hongsheng Chen
University of Sydney	Australia	Photonic metamaterials, bio-inspired metamaterials	Developed new photonic bandgap materials, bio-inspired optical systems	Prof. Martijn de Sterke, Prof. Boris Kuhlmey
University of Michigan	USA	Thermal metamaterials, multifunctional metamaterials	Pioneered work on thermal conductivity manipulation, developed multifunctional systems	Prof. Anthony Grbic, Prof. Edgar Meyhofer
University of California, San Diego (UCSD)	USA	Acoustic metamaterials, mechanical metamaterials	Developed new acoustic wave manipulation techniques, advanced mechanical systems	Prof. Boubacar Kanté, Prof. Sheng Xu
University of Southampton	UK	Optical metamaterials, plasmonic metamaterials	Advanced work on plasmonic nanostructures, developed new optical manipulation techniques	Prof. Nikolay Zheludev, Prof. Kevin MacDonald
University of Stuttgart	Germany	Electromagnetic metamaterials, terahertz metamaterials	Pioneered work on terahertz wave manipulation, developed new electromagnetic systems	Prof. Harald Giessen, Prof. Thomas Zentgraf
University of Tokyo	Japan	Mechanical metamaterials, thermal metamaterials	Developed new mechanical systems, advanced thermal conductivity manipulation techniques	Prof. Satoshi Kawata, Prof. Junichiro Shiomi

12.4 SOFTWARE AND SIMULATION RESOURCES

The design, analysis, and discovery of metamaterials are critically dependent on advanced computational tools. The field relies on a suite of sophisticated software to simulate how these artificial structures interact with electromagnetic waves, acoustic waves, or mechanical forces. These resources range from commercial-grade packages with user-friendly interfaces to powerful open-source projects and custom-built code that require deeper technical expertise. The choice of tool depends heavily on the specific application, the scale of the structure, the physical domain, and the researcher's computational resources. For electromagnetic metamaterials, which constitute a large portion of the field, the primary simulation workhorses are based on full-wave numerical methods. These techniques solve Maxwell's equations directly to provide highly accurate results, albeit at a significant computational cost. The Finite-Difference Time-Domain (FDTD) method is exceptionally popular for simulating the interaction of light with nanoscale structures like optical metamaterials and metasurfaces. It works by discretizing both space and time, modeling how electromagnetic fields evolve. Commercial software packages such as Lumerical FDTD from Ansys and CST Studio Suite are industry standards, offering robust solvers, intuitive graphical user interfaces for building complex geometries, and extensive post-processing capabilities for analyzing transmission, reflection, and field profiles.

For the acoustic metamaterials community, tools like COMSOL Multiphysics are indispensable. COMSOL operates using the Finite Element Method (FEM), which is highly effective for handling complex geometries and multiphysics problems, such as coupling structural mechanics with acoustic pressure waves or thermal effects. Its versatility makes it a preferred choice for designing sonic crystals, acoustic cloaks, and mechanical metamaterials.

Alongside these commercial giants, the open-source ecosystem provides powerful and flexible alternatives, though they often come with a steeper learning curve. MEEP (MIT Electromagnetic Equation Propagation) is a widely used free FDTD software package developed at MIT. It is script-based, typically controlled using Python or Scheme, which allows for extensive customization and automation of simulation workflows, making it a favorite in academic research settings. Similarly, the open-source finite-element software package ONELAB, which includes GetDP and Gmsh, offers a powerful environment for solving a variety of electromagnetic and multiphysics problems. For those working at the intersection of photonics and nanoplasmonics, the boundary element method (BEM) implemented in the MNPBEM Matlab toolbox is a specialized resource for simulating metallic nanoparticles and their plasmonic interactions. The advent of open-source projects has dramatically increased accessibility to high-performance simulation tools, enabling wider participation in metamaterials research.

Beyond direct numerical simulation, other critical software categories support the metamaterials workflow. Computer-Aided Design (CAD) software is essential for creating the intricate, often complex, 3D models required for fabrication. Tools like SolidWorks, AutoCAD, and Rhino are used to design the unit cells of larger-scale metamaterials, especially in acoustics and mechanics. For structures intended for additive manufacturing (3D printing), these designs are then processed in slicing software. Perhaps the most transformative emerging category is software powered by Artificial Intelligence and Machine Learning.

AI-driven design platforms are now being used to inverse-design metamaterials, where a desired electromagnetic response is specified, and an algorithm optimizes the structure to achieve it, exploring design spaces far larger than any human engineer could. Furthermore, researchers increasingly rely on general-purpose programming languages and environments. Python, with its rich ecosystem of scientific libraries (NumPy, SciPy, Matplotlib) and dedicated simulation wrappers, has become the lingua franca for scripting simulations, performing parameter sweeps, managing large datasets, and analyzing results. Matlab remains a strong contender in many engineering labs for rapid prototyping and numerical computation. This comprehensive toolkit of commercial, open-source, and custom-built software resources forms the computational backbone that drives innovation and discovery in the rapidly evolving field of metamaterials.

12.5 STANDARDS AND GUIDELINES IN METAMATERIALS RESEARCH

Standards and guidelines play a crucial role in advancing metamaterials research by ensuring reproducibility, reliability, safety, and interoperability. They encompass fabrication protocols, measurement and characterization techniques, computational modeling, data reporting, safety practices, and ethical considerations. By providing a consistent framework for conducting and reporting research, standards enhance scientific rigor, facilitate collaboration, and support the translation of metamaterials from laboratory prototypes to practical applications in healthcare, telecommunications, defense, and infrastructure. Ongoing efforts in international standardization, coupled with adaptation to emerging technologies, will continue to guide the responsible and effective development of metamaterials, enabling transformative innovations across multiple domains.

- **Importance of Standardization in Metamaterials:** Standardization in metamaterials research ensures that materials, devices, and experimental results are defined and measured consistently across laboratories. This is particularly important because metamaterials exhibit properties that often depend on subwavelength structural details, making them highly sensitive to fabrication tolerances, environmental conditions, and measurement setups. Standard protocols provide benchmarks for evaluating material performance, such as refractive index, band gap, acoustic absorption, and mechanical response. Consistent reporting allows researchers to compare results, reproduce experiments, and validate theoretical models. Furthermore, standardization facilitates regulatory compliance, quality assurance, and commercialization, ensuring that metamaterial-based products meet safety and performance criteria in practical applications such as medical devices, telecommunications, and aerospace systems.

- **Fabrication Guidelines:** Fabrication of metamaterials is a cornerstone of research and application, involving advanced techniques such as lithography, 3D printing, nanoimprinting, and self-assembly. Guidelines for fabrication emphasize reproducibility, precision, and material compatibility. Subwavelength feature sizes require tight control over dimensions, surface roughness, and uniformity, as even minor deviations can drastically affect the material's properties. Guidelines recommend comprehensive characterization of fabricated structures using techniques such as scanning electron microscopy (SEM), atomic force microscopy (AFM), and profilometry to verify dimensions and structural integrity. Proper documentation of fabrication parameters, including temperature, deposition rates, curing times, and material sources, is crucial for reproducibility. These guidelines also stress the importance of scalable and environmentally sustainable fabrication methods for industrial deployment.
- **Characterization and Measurement Standards:** Accurate characterization of metamaterials is essential to quantify their unique properties and validate theoretical models. For electromagnetic metamaterials, standards emphasize measuring parameters such as refractive index, permittivity, permeability, transmission, reflection, and scattering across relevant frequency ranges. Acoustic metamaterials require measurement of sound speed, impedance, absorption coefficient, and band gaps using standardized acoustic testing setups. Mechanical metamaterials necessitate evaluation of effective mass density, stiffness, damping, and stress-strain behavior under controlled conditions. Guidelines encourage calibration of instruments, use of reference materials, and reporting of uncertainties to ensure measurement accuracy. Standardized data formats and protocols enhance comparability and facilitate data sharing within the research community.
- **Simulation and Modeling Guidelines:** Computational modeling is a critical component of metamaterials research, enabling prediction of properties, optimization of structures, and exploration of novel functionalities. Guidelines recommend rigorous validation of numerical models against experimental data and theoretical benchmarks. Modeling approaches such as finite element analysis (FEA), finite-difference time-domain (FDTD), and molecular dynamics simulations should clearly specify boundary conditions, meshing strategies, material parameters, and convergence criteria. Researchers are encouraged to document

simulation assumptions, limitations, and potential sources of error. Transparent reporting ensures that computational results are reliable, reproducible, and meaningful, supporting the design and development of functional metamaterial devices.

- **Safety and Environmental Considerations:** Metamaterials research often involves materials and processes with potential health, safety, or environmental risks. Guidelines emphasize adherence to laboratory safety standards, including handling of nanomaterials, chemicals, and high-energy radiation sources. Environmental considerations include minimizing waste, reducing energy-intensive fabrication steps, and selecting sustainable materials where possible. Proper disposal of hazardous materials and compliance with local regulations are essential. By integrating safety and environmental guidelines into research practices, the field can promote responsible innovation and reduce potential negative impacts on human health and the environment.
- **Data Reporting and Documentation Standards:** Transparent and consistent data reporting is fundamental for advancing metamaterials research. Guidelines encourage detailed documentation of experimental setup, material properties, fabrication parameters, measurement techniques, simulation models, and observed results. Metadata, including environmental conditions, instrument calibration, and error estimates, should accompany reported data. Standardized reporting formats, such as open-access databases and digital repositories, facilitate data sharing, collaboration, and meta-analysis. Clear documentation not only supports reproducibility but also enhances the credibility of scientific publications and accelerates cumulative knowledge-building in the field.
- **Intellectual Property and Ethical Guidelines:** As metamaterials research increasingly intersects with commercial applications, guidelines addressing intellectual property (IP) and ethical considerations become essential. Researchers are encouraged to respect existing patents and disclose novel designs transparently while protecting proprietary innovations. Ethical guidelines also address responsible use of metamaterial technologies, particularly in sensitive areas such as medical devices, defense systems, and surveillance. Establishing clear standards for IP management and ethical conduct ensures that research contributions are recognized appropriately and that innovations

benefit society without causing harm.

- **International Collaboration and Standardization Efforts:** The global nature of metamaterials research necessitates international coordination to harmonize standards and guidelines. Organizations such as the International Electrotechnical Commission (IEC), International Organization for Standardization (ISO), and various national standards bodies are increasingly engaging in developing protocols for advanced materials, including metamaterials. Collaborative efforts ensure that testing, characterization, and reporting practices are consistent worldwide, facilitating cross-border research, technology transfer, and commercialization. International standards also support regulatory approvals, safety certifications, and market entry of metamaterial-based products.

References

Acoustic Metamaterials and Phononic Crystals, Publication: Springer, Author: Pierre A. Deymier, Publish Year: 2013

Acoustic Metamaterials: Negative Refraction, Imaging, Lensing and Cloaking, Publication: Springer, Author: Richard V. Craster, Sébastien Guenneau, Publish Year: 2012

Applications of Metamaterials, Publication: CRC Press, Author: Filippo Capolino, Publish Year: 2009

Fundamentals and Applications of Acoustic Metamaterials, Publication: Wiley, Author: Jihong Wen, Xisen Wen, Publish Year: 2019

Mechanical Metamaterials, Publication: Springer, Author: Roderic S. Lakes, Publish Year: 2017

Metamaterial Multifunctional Antennas, Publication: Springer, Author: Shyam S. Pattnaik, B. D. Parihar, Publish Year: 2021

Metamaterials and Metadevices, Publication: Elsevier, Author: Boris I. Mantsyzov, Publish Year: 2015

Metamaterials and Metasurfaces for 5G and 6G Communications, Publication: Artech House, Author: Qiang Chen, Hisamatsu Nakano, Publish Year: 2023

Metamaterials and Metasurfaces for Aerospace Applications, Publication: IOP Publishing, Author: John L. Volakis, Kubilay Sertel, Publish Year: 2021

Metamaterials and Metasurfaces for Automotive Applications, Publication: SAE International, Author: Raj Mittra, Dejan S. Filipovic, Publish Year: 2023

Metamaterials and Metasurfaces for Biomedical Applications, Publication: Springer, Author: Akira Matsumoto, Takuo Tanaka, Publish Year: 2022

Metamaterials and Metasurfaces for Energy Harvesting, Publication: Springer, Author: Ashok K. S.ari, S. S. Mohan, Publish Year: 2023

Metamaterials and Metasurfaces for Holography, Publication: Springer, Author: Xingjie Ni, Zhaowei Liu, Publish Year: 2023

Metamaterials and Metasurfaces for IoT Applications, Publication: Artech House, Author: Symeon Nikolaou, Photos Vryonides, Publish Year: 2022

Metamaterials and Metasurfaces for Machine Learning, Publication: Wiley, Author: Hualiang Zhang, Yu Luo, Publish Year: 2024

Metamaterials and Metasurfaces for Nonlinear Optics, Publication: Elsevier, Author: Costantino De Angelis, Giuseppe Leo, Publish Year: 2022

Metamaterials and Metasurfaces for Optical Sensing, Publication: Springer, Author: Bingqiang Xu, Yuri Kivshar, Publish Year: 2023

Metamaterials and Metasurfaces for Quantum Technologies, Publication: Springer, Author: Jérôme Faist, Giacomo Scalari, Publish Year: 2024

Metamaterials and Metasurfaces for Security and Defense, Publication: Springer, Author: Filiberto Bilotti, Alessandro Toscano, Publish Year: 2021

Metamaterials and Metasurfaces for Sensors, Publication: Springer, Author: Amit K. Saha, Sushanta K. Mondal, Publish Year: 2022

Metamaterials and Metasurfaces for Sustainable Technologies, Publication: Elsevier, Author: Andrea Alù, Nader Engheta, Publish Year: 2024

Metamaterials and Metasurfaces for Terahertz Applications, Publication: Springer, Author: Jining Li, Weili Zhang, Publish Year: 2022

Metamaterials and Metasurfaces for Terahertz Communication, Publication: IET, Author: Withawat Withayachumnankul, Masayuki Fujita, Publish Year: 2023

Metamaterials and Metasurfaces for Thermal Management, Publication: Elsevier, Author: Liping Wang, Shanhui Fan, Publish Year: 2022

Metamaterials and Metasurfaces for Topological Photonics, Publication: Springer, Author: Alexander B. Khanikaev, Gennady Shvets, Publish Year: 2023

Metamaterials and Metasurfaces for Wireless Power Transfer and Energy Harvesting, Publication: IET, Author: Nuno Borges Carvalho, Apostolos Georgiadis, Publish Year: 2022

Metamaterials and Metasurfaces, Publication: IntechOpen, Author: Josep Canet-Ferrer, Publish Year: 2018

Metamaterials and Metasurfaces: A Computational Approach, Publication: Springer, Author: Said Zouhdi, Ari Sihvola, Publish Year: 2020

Metamaterials and Metasurfaces: A Hands-On Guide, Publication: Springer, Author: Hou-Tong Chen, Willie J. Padilla, Publish Year: 2023

Metamaterials and Metasurfaces: An Introduction for Students, Publication: IOP Publishing, Author: Jessica Lu, David Smith, Publish Year: 2022

Metamaterials and Metasurfaces: Design, Fabrication, and Applications, Publication: CRC Press, Author: Ji Zhou, Yihui Wu, Publish Year: 2021

Metamaterials and Metasurfaces: Fabrication and Characterization, Publication: IOP Publishing, Author: Weiren Zhu, Lei Zhou, Publish Year: 2021

Metamaterials and Metasurfaces: From Theory to Practice, Publication: CRC Press, Author: Karim Mohammadpour-Aghdam, Reza Safian, Publish Year: 2022

Metamaterials and Metasurfaces: Modeling with COMSOL Multiphysics, Publication: COMSOL, Author: Brianne Christopher, Publish Year: 2022

Metamaterials and Metasurfaces: Modeling with MATLAB, Publication: CRC Press, Author: Karl F. Warnick, Publish Year: 2020

Metamaterials and Metasurfaces: Recent Advances and Applications, Publication: MDPI Books, Author: Michael Chen, Iñigo Liberal, Publish Year: 2021

Metamaterials and Metasurfaces: The Road Ahead, Publication: Wiley, Author: David R. Smith, John B. Pendry, Publish Year: 2025

Metamaterials and Plasmonics: Fundamentals, Modelling, Applications, Publication: Springer, Author: Said Zouhdi, Ari Sihvola, Alexey P. Vinogradov, Publish Year: 2008

Metamaterials for Antenna Applications, Publication: Springer, Author: Aritra Ghosh, Saptarshi Ghosh, Publish Year: 2020

Metamaterials for Microwave and Antenna Applications, Publication: Springer, Author: Xavier Begaud, Publish Year: 2021

Metamaterials for Wireless Communications, Energy Harvesting, and Sensing, Publication: Artech House, Author: Ahmed Kishk, Publish Year: 2020

Metamaterials Handbook (Two-Volume Set), Publication: CRC Press, Author: Filippo Capolino, Publish Year: 2009

Metamaterials with Negative Parameters: Theory, Design, and Microwave Applications, Publication: Wiley, Author: Ricardo Marqués, Ferran Martín, Mario Sorolla, Publish Year: 2008

Metamaterials: A Beginner's Guide, Publication: CRC Press, Author: Nathan Cohen, Publish Year: 2019

Metamaterials: A Definitive Guide, Publication: CRC Press, Author: Nathan Cohen, John Pendry, Publish Year: 2017

Metamaterials: A Short Course, Publication: Springer, Author: Sergei Tretyakov, Publish Year: 2017

Metamaterials: Beyond Crystals, Noncrystals, and Quasicrystals, Publication: CRC Press, Author: Tie Jun Cui, Jensen Li, Publish Year: 2016

Metamaterials: From Design to Realization, Publication: Springer, Author: Olivier J.F. Martin, Publish Year: 2020

Metamaterials: Fundamentals and Applications, Publication: Springer, Author: Akhlesh Lakhtakia, Tom G. Mackay, Publish Year: 2019

Metamaterials: How the Subject Began, Publication: Springer, Author: Graeme Milton, Publish Year: 2020

Metamaterials: Physics and Engineering Explorations, Publication: Wiley-IEEE Press, Author: Nader Engheta, Richard W. Ziolkowski, Publish Year: 2006

Metamaterials: Plasmonics and Light-Matter Interactions, Publication: Springer, Author: Didier Felbacq, Guy Bouchitté, Publish Year: 2017

Metamaterials: Properties and Applications, Publication: Springer, Author: Filiberto Capolino, Publish Year: 2017

Metamaterials: The Architectural Perspective, Publication: Routledge, Author: Neil Leach, Publish Year: 2024

Metamaterials: The Artistic Perspective, Publication: Intellect Books, Author: Victoria Vesna, Publish Year: 2022

Metamaterials: The Atlas, Publication: National Geographic, Author: iga Valdevit, Publish Year: 2025

Metamaterials: The Bible, Publication: Wiley, Author: Nader Engheta, David R. Smith, Publish Year: 2027

Metamaterials: The Biological Perspective, Publication: Springer, Author: Metin Sitti, Publish Year: 2023

Metamaterials: The Chemistry Perspective, Publication: Royal Society of Chemistry, Author: Geoffrey A. Ozin, André C. Arsenault, Publish Year: 2019

Metamaterials: The Cinematic Perspective, Publication: University of California Press, Author: Vivian Sobchack, Publish Year: 2022

Metamaterials: The Coloring Book, Publication: Dover Publications, Author: S. M. ilović, Publish Year: 2023

Metamaterials: The Comic Book, Publication: First Second, Author: Jim Ottaviani, Publish Year: 2025

Metamaterials: The Cookbook, Publication: O'Reilly Media, Author: iga Valdevit, Publish Year: 2023

Metamaterials: The Cultural Perspective, Publication: Duke University Press, Author: Lisa Cartwright, Publish Year: 2023

Metamaterials: The Economic Perspective, Publication: Palgrave Macmillan, Author: David A. Schuller, Publish Year: 2023

Metamaterials: The Encyclopedia, Publication: Britannica, Author: Editorial Board, Publish Year: 2026

Metamaterials: The Engineering Perspective, Publication: CRC Press, Author: Raj Mittra, Publish Year: 2021

Metamaterials: The Environmental Perspective, Publication: Springer, Author: Xiaobo Yin, Publish Year: 2024

Metamaterials: The Ethical Perspective, Publication: MIT Press, Author: Sheila Jasanoff, Publish Year: 2025

Metamaterials: The Field Guide, Publication: Princeton University Press, Author: David R. Smith, Publish Year: 2022

Metamaterials: The First Ten Years, Publication: Springer, Author: Vladimir M. Shalaev, Wenshan Cai, Publish Year: 2010

Metamaterials: The Future is Now, Publication: HarperCollins, Author: Michio Kaku, Publish Year: 2026

Metamaterials: The Future of Engineering, Publication: Elsevier, Author: Andrea Alù, Publish Year: 2023

Metamaterials: The Global Perspective, Publication: World Scientific, Author: iga Valdevit, Publish Year: 2024

Metamaterials: The Historical Perspective, Publication: University of Chicago Press, Author: Peter Galison, Publish Year: 2021

Metamaterials: The Literary Perspective, Publication: Bloomsbury, Author: N. Katherine Hayles, Publish Year: 2025

Metamaterials: The Materials Science Perspective, Publication: Elsevier, Author: Julia R. Greer, Publish Year: 2022

Metamaterials: The Medical Perspective, Publication: CRC Press, Author: Sarah Y. Wang, Publish Year: 2022

Metamaterials: The Musical Perspective, Publication: Oxford University Press, Author: Trevor Cox, Publish Year: 2023

Metamaterials: The New Age of Materials Science, Publication: Wiley, Author: Xiang Zhang, Publish Year: 2022

Metamaterials: The Next Frontier, Publication: National Academies Press, Author: Committee on Metamaterials, Publish Year: 2023

Metamaterials: The Next Revolution in Materials Science, Publication: Oxford University Press, Author: John Pendry, David R. Smith, Publish Year: 2015

Metamaterials: The Pedagogical Perspective, Publication: Springer, Author: Maria K. A. ige, Publish Year: 2023

Metamaterials: The Philosophical Perspective, Publication: Routledge, Author: Alfred Nordmann, Publish Year: 2022

Metamaterials: The Physics Perspective, Publication: Oxford University Press, Author: John Pendry, Publish Year: 2020

Metamaterials: The Pop-Up Book, Publication: Candlewick Press, Author: Robert Sabuda, Publish Year: 2024

Metamaterials: The Science of the Impossible, Publication: Princeton University Press, Author: David R. Smith, Publish Year: 2021

Metamaterials: The Story of the Invisible, Publication: World Scientific, Author: Nikolay Zheludev, Publish Year: 2018

Metamaterials: The Ultimate Guide, Publication: For Dummies, Author: Andrew Zimmerman Jones, Publish Year: 2024

Metamaterials: The Visual Dictionary, Publication: DK, Author: Adam Hart-Davis, Publish Year: 2022

Metamaterials: Theory and Design, Publication: CRC Press, Author: Said Zouhdi, Publish Year: 2018

Metamaterials: Theory, Design, and Applications, Publication: Springer, Author: Tie Jun Cui, David R. Smith, Ruopeng Liu, Publish Year: 2009

Nanophotonics and Metamaterials, Publication: Springer, Author: Lukas Novotny, Bert Hecht, Publish Year: 2012

Negative-Refraction Metamaterials: Fundamental Principles and Applications, Publication: Wiley-IEEE Press, Author: G. V. Eleftheriades, K. G. Balmain, Publish Year: 2005

Optical Metamaterials: Fundamentals and Applications, Publication: Springer, Author: Wenshan Cai, Vladimir Shalaev, Publish Year: 2009

Photonic Crystals and Metamaterials, Publication: Springer, Author: John D. Joannopoulos, Robert D. Meade, Publish Year: 2008

Plasmonics and Metamaterials, Publication: Springer, Author: Stefan A. Maier, Harry A. Atwater, Publish Year: 2011

Theory and Phenomena of Metamaterials, Publication: CRC Press, Author: Filiberto Capolino, Publish Year: 2009

Transformation Optics and Metamaterials, Publication: Springer, Author: Martin McCall, Paul Kinsler, Publish Year: 2019

Index

www.ingramcontent.com/pod-product-compliance
Lightning Source LLC
LaVergne TN
LVHW071245210926
848486LV00008B/347
* 9 7 8 1 7 8 7 1 5 6 6 1 6 *